T0137433

Studies in Big Data

Volume 43

Series editor

Janusz Kacprzyk, Polish Academy of Sciences, Warsaw, Poland
e-mail: kacprzyk@ibspan.waw.pl

The series "Studies in Big Data" (SBD) publishes new developments and advances in the various areas of Big Data- quickly and with a high quality. The intent is to cover the theory, research, development, and applications of Big Data, as embedded in the fields of engineering, computer science, physics, economics and life sciences. The books of the series refer to the analysis and understanding of large, complex, and/or distributed data sets generated from recent digital sources coming from sensors or other physical instruments as well as simulations, crowd sourcing, social networks or other internet transactions, such as emails or video click streams and others. The series contains monographs, lecture notes and edited volumes in Big Data spanning the areas of computational intelligence including neural networks, evolutionary computation, soft computing, fuzzy systems, as well as artificial intelligence, data mining, modern statistics and operations research, as well as self-organizing systems. Of particular value to both the contributors and the readership are the short publication timeframe and the world-wide distribution, which enable both wide and rapid dissemination of research output.

More information about this series at http://www.springer.com/series/11970

Mamta Mittal · Valentina E. Balas
Lalit Mohan Goyal · Raghvendra Kumar
Editors

Big Data Processing Using Spark in Cloud

Springer

Editors
Mamta Mittal
Department of Computer Science
  and Engineering
GB Pant Government Engineering College
New Delhi
India

Valentina E. Balas
Department of Automation
  and Applied Informatics
Aurel Vlaicu University of Arad
Arad
Romania

Lalit Mohan Goyal
Department of Computer Science
  and Engineering
Bharati Vidyapeeth's College of
  Engineering
New Delhi
India

Raghvendra Kumar
Department of Computer Science
  and Engineering
Laxmi Narayan College of Technology
Jabalpur, Madhya Pradesh
India

ISSN 2197-6503   ISSN 2197-6511 (electronic)
Studies in Big Data
ISBN 978-981-13-4448-0   ISBN 978-981-13-0550-4 (eBook)
https://doi.org/10.1007/978-981-13-0550-4

This Springer imprint is published by the registered company Springer Nature Singapore Pte Ltd.
The registered company address is: 152 Beach Road, #21-01/04 Gateway East, Singapore 189721, Singapore

Preface

The edited book "Big Data Processing using Spark in Cloud" takes deep into Spark while starting with the basics of Scala and core Spark framework, and then explore Spark data frames, machine learning using MLlib, graph analytics using graph X, and real-time processing with Apache Kafka, AWS Kinesis, and Azure Event Hub. We will also explore Spark using PySpark and R., apply the knowledge that so far we have learnt about Spark, and will work on real datasets and do some exploratory analytics first, then move on to predictive modeling on Boston Housing Datasets, and then move forward to build news content-based recommender system using NLP and MLlib, collaborative filtering-based movies recommender system, and page rank using GraphX. This book also discusses how to tune Spark parameters for production scenarios and how to write robust applications in Apache Spark using Scala in cloud computing environment.

The book is organized into 11 chapters.

Chapter "A Survey on Big Data—Its Challenges and Solution from Vendors" carried out a detailed survey depicting the enormous information and its difficulties alongside the advancements required to deal with huge data. This moreover portrays the conventional methodologies which were utilized before to manage information, their impediments, and how it is being overseen by the new approach Hadoop. It additionally portrays the working of Hadoop along with its pros and cons and security on huge data.

Chapter "Big Data Streaming with Spark" introduces many concepts associated with Spark Streaming, including a discussion of supported operations. Finally, two other important platforms and their integration with Spark, namely Apache Kafka and Amazon Kinesis, are explored.

Chapter "Big Data Analysis in Cloud and Machine Learning" discusses data which is considered to be the lifeblood of any business organization, as it is the data that streams into actionable insights of businesses. The data available with the organizations is so much in volume that it is popularly referred as Big Data. It is the hottest buzzword spanning the business and technology worlds. Economies over the world are using Big Data and Big Data analytics as a new frontier for business so as to plan smarter business moves, improve productivity and performance, and plan strategy more effectively. To make Big Data analytics effective, storage technologies and analytical tools play a critical role. However, it is evident that Big Data places rigorous demands on networks, storage, and servers, which has motivated organizations and enterprises to move on cloud, in order to harvest maximum benefits of the available Big Data. Furthermore, we are also aware that traditional analytics tools are not well suited to capturing the full value of Big Data. Hence, machine learning seems to be an ideal solution for exploiting the opportunities hidden in Big Data. In this chapter, we shall discuss Big Data and Big Data analytics with a special focus on cloud computing and machine learning.

Chapter "Cloud Computing Based Knowledge Mapping Between Existing and Possible Academic Innovations—*An Indian Techno-Educational Context*" discusses various applications in cloud computing that allow healthy and wider efficient computing services in terms of providing centralized services of storage, applications, operating systems, processing, and bandwidth. Cloud computing is a type of architecture which helps in the promotion of scalable computing. Cloud computing is also a kind of resource-sharing platform and thus needed in almost all the spectrum and areas regardless of its type. Today, cloud computing has a wider market, and it is growing rapidly. The manpower in this field is mainly outsourced from the IT and computing services, but there is an urgent need to offer cloud computing as full-fledged bachelors and masters programs. In India also, cloud computing is rarely seen as an education program, but the situation is now changing. There is high potential to offer cloud computing in Indian educational segment. This paper is conceptual in nature and deals with the basics of cloud computing, its need, features, types existing, and possible programs in the Indian context, and also proposed several programs which ultimately may be helpful for building solid Digital India.

The objective of the Chapter "Data Processing Framework Using Apache and Spark Technologies in Big Data" is to provide an overall view of Hadoop's MapReduce technology used for batch processing in cluster computing. Then, Spark was introduced to help Hadoop work faster, but it can also work as a stand-alone system with its own processing engine that uses Hadoop's distributed file storage or cloud storage of data. Spark provides various APIs according to the type of data and processing required. Apart from that, it also provides tools for query processing, graph processing, and machine learning algorithms. Spark SQL is a very important framework of Spark for query processing and maintains storage of large datasets on cloud. It also allows taking input data from different data sources and performing operations on it. It provides various inbuilt functions to directly create and maintain data frames.

Chapter "Implementing Big Data Analytics Through Network Analysis Software Applications in Strategizing Higher Learning Institutions" discusses the common utility among these social media applications, so that they are able to create natural network data. These online social media networks (OSMNs) represent the links or relationships between content generators as they look, react, comment, or link to one another's content. There are many forms of computer-mediated social interaction which includes SMS messages, emails, discussion groups, blogs, wikis, videos, and photograph-sharing systems, chat rooms, and "social network services." All these applications generate social media datasets of social friendships. Thus OSMNs have academic and pragmatic value and can be leveraged to illustrate the crucial contributors and the content. Our study considered all the above points into account and explored the various Network Analysis Software Applications to study the practical aspects of Big Data analytics that can be used to better strategies in higher learning institutions.

Chapter "Machine Learning on Big Data: A Developmental Approach on Societal Applications" concentrates on the most recent progress over researches with respect to machine learning for Big Data analytic and different techniques in the context of modern computing environments for various societal applications. Specifically, our aim is to investigate the opportunities and challenges of ML on Big Data and how it affects the society. The chapter covers a discussion on ML in Big Data in specific societal areas.

Chapter "Personalized Diabetes Analysis Using Correlation Based Incremental Clustering Algorithm" describes the details about incremental clustering approach, correlation-based incremental clustering algorithm (CBICA) to create clusters by applying CBICA to the data of diabetic patients and observing any relationship which indicates the reason behind the increase of the diabetic level over a specific period of time including frequent visits to healthcare facility. These obtained results from CBICA are compared with the results obtained from other incremental clustering approaches, closeness factor-based algorithm (CFBA), which is a probability-based incremental clustering algorithm. "Cluster-first approach" is the distinctive concept implemented in both CFBA and CBICA algorithms. Both these algorithms are "parameter-free," meaning only end user requires to give input dataset to these algorithms, and clustering is automatically performed using no additional dependencies from user including distance measures, assumption of centroids, and number of clusters to form. This research introduces a new definition of outliers, ranking of clusters, and ranking of principal components.

Scalability: Such personalization approach can be further extended to cater the needs of gestational, juvenile, and type 1 and type 2 diabetic prevention in society. Such research can be further made distributed in nature so as to consider diabetic patients' data from all across the world and for wider analysis. Such analysis may vary or can be clustered based on seasonality, food intake, personal exercise regime, heredity, and other related factors.

Without such integrated tool, the diabetologist in hurry, while prescribing new details, may consider only the latest reports, without empirical details of an individual. Such situations are very common in these stressful and time-constraint lives, which may affect the accurate predictive analysis required for the patient.

Chapter "Processing Using Spark—A Potent of BD Technology" sustains the major potent of processing behind Spark-connected contents like resilient distributed datasets (RDDs), scalable machine learning libraries (MLlib), Spark incremental streaming pipeline process, parallel graph computation interface through GraphX, SQL data frames, Spark SQL (data processing paradigm supports columnar storage), and recommendation systems with MlLib. All libraries operate on RDDs as the data abstraction is very easy to compose with any applications. RDDs are fault-tolerant computing engines (RDDs are the major abstraction and provide explicit support for data sharing (user's computations) and can capture a wide range of processing workloads and fault-tolerant collections of objects partitioned across a cluster which can be manipulated in parallel). These are exposed through functional programming APIs (or BD supported languages) like Scala and Python. This chapter also throws a viewpoint on core scalability of Spark to build high-level data processing libraries for the next generation of computer applications, wherein a complex sequence of processing steps is involved. To understand and simplify the entire BD tasks, focusing on the processing hindsight, insights, foresight by using Spark's core engine, its members of ecosystem components are explained with a neat interpretable way, which is mandatory for data science compilers at this moment. One of the tools in Spark, cloud storage, is explored in this initiative to replace the bottlenecks toward the development of an efficient and comprehend analytics applications.

Chapter "Recent Developments in Big Data Analysis Tools and Apache Spark" illustrates different tools used for the analysis of Big Data in general and Apache Spark (AS) in particular. The data structure used in AS is Spark RDD, and it also uses Hadoop. This chapter also entails merits, demerits, and different components of AS tool.

Chapter "SCSI: Real-Time Data Analysis with Cassandra and Spark" focused on understanding the performance evaluations, and Smart Cassandra Spark Integration (SCSI) streaming framework is compared with the file system-based data stores such as Hadoop streaming framework. SCSI framework is found scalable, efficient, and accurate while computing big streams of IoT data.

There have been several influences from our family and friends who have sacrificed a lot of their time and attention to ensure that we are kept motivated to complete this crucial project.

The editors are thankful to all the members of Springer (India) Private Limited, especially Aninda Bose and Jennifer Sweety Johnson for the given opportunity to edit this book.

New Delhi, India   Mamta Mittal
Arad, Romania   Valentina E. Balas
New Delhi, India   Lalit Mohan Goyal
Jabalpur, India   Raghvendra Kumar

Contents

About the Editors

Mamta Mittal, Ph.D. is working in GB Pant Government Engineering College, Okhla, New Delhi. She graduated in Computer Science and Engineering from Kurukshetra University, Kurukshetra, and received masters' degree (Honors) in Computer Science and Engineering from YMCA, Faridabad. She has completed her Ph.D. in Computer Science and Engineering from Thapar University, Patiala. Her research area includes data mining, Big Data, and machine learning algorithms. She has been teaching for last 15 years with an emphasis on data mining, DBMS, operating system, and data structure. She is Active Member of CSI and IEEE. She has published and communicated a number of research papers and attended many workshops, FDPs, and seminars as well as one patent (CBR no. 35107, Application number: 201611039031, a semiautomated surveillance system through fluoroscopy using AI techniques). Presently, she is supervising many graduates, postgraduates, and Ph.D. students.

Valentina E. Balas, Ph.D. is currently Full Professor in the Department of Automatics and Applied Software at the Faculty of Engineering, "Aurel Vlaicu" University of Arad, Romania. She holds a Ph.D. in Applied Electronics and Telecommunications from Polytechnic University of Timisoara. She is author of more than 270 research papers in refereed journals and international conferences. Her research interests are in intelligent systems, fuzzy control, soft computing, smart sensors, information fusion, modeling and simulation. She is the Editor-in-Chief to International Journal of Advanced Intelligence Paradigms (IJAIP) and to International Journal of Computational Systems Engineering (IJCSysE), Member in Editorial Board member of several national and international journals and is evaluator expert for national and international projects. She served as General Chair of the International Workshop Soft Computing and Applications in seven editions 2005–2016 held in Romania and Hungary. She participated in many international conferences as Organizer, Session Chair, and Member in International Program Committee. Now she is working in a national project with EU funding support: BioCell-NanoART = Novel Bio-inspired Cellular Nano-Architectures—For Digital Integrated Circuits, 2M Euro from National

Authority for Scientific Research and Innovation. She is a Member of EUSFLAT, ACM, and a Senior Member IEEE, Member in TC—Fuzzy Systems (IEEE CIS), Member in TC—Emergent Technologies (IEEE CIS), Member in TC—Soft Computing (IEEE SMCS). She was Vice President (Awards) of IFSA International Fuzzy Systems Association Council (2013–2015) and is a Joint Secretary of the Governing Council of Forum for Interdisciplinary Mathematics (FIM)—A Multidisciplinary Academic Body, India.

Lalit Mohan Goyal, Ph.D. has completed Ph.D. from Jamia Millia Islamia, New Delhi, in Computer Engineering, M.Tech. (Honors) in Information Technology from Guru Gobind Singh Indraprastha University, New Delhi, and B.Tech. (Honors) in Computer Science and Engineering from Kurukshetra University, Kurukshetra. He has 14 years of teaching experience in the area of parallel and random algorithms, data mining, cloud computing, data structure, and theory of computation. He has published and communicated a number of research papers and attended many workshops, FDPs, and seminars. He is a reviewer for many reputed journals. Presently, he is working at Bharti Vidyapeeth's College of Engineering, New Delhi.

Raghvendra Kumar, Ph.D. has been working as Assistant Professor in the Department of Computer Science and Engineering at LNCT College, Jabalpur, MP, and as a Ph.D. (Faculty of Engineering and Technology) at Jodhpur National University, Jodhpur, Rajasthan, India. He completed his Master of Technology from KIIT University, Bhubaneswar, Odisha, and his Bachelor of Technology from SRM University, Chennai, India. His research interests include graph theory, discrete mathematics, robotics, cloud computing and algorithm. He also works as a reviewer and an editorial and technical board member for many journals and conferences. He regularly publishes research papers in international journals and conferences and is supervising postgraduate students in their research work.

Key Features

1. Covers all the Big Data analysis using Spark
2. Covers the complete data science workflow in cloud
3. Covers the basics and high-level concepts, thus serves as a cookbook for industry persons and also helps beginners to learn things from basic to advance
4. Covers privacy issue and challenges for Big Data analysis in cloud computing environment
5. Covers the major changes and advancement of Big Data analysis
6. Covers the concept of Big Data analysis technologies and their applications in real world
7. Data processing, analysis, and security solutions in cloud environment.

A Survey on Big Data—Its Challenges and Solution from Vendors

Kamalinder Kaur and Vishal Bharti

Abstract Step by step there comes another innovation, gadgets and techniques which offer ascent to the fast development of information. Presently today, information is immensely expanding inside each ten minutes and it is difficult to oversee it and it offers ascend to the term Big data. This paper depicts the enormous information and its difficulties alongside the advancements required to deal with huge data. This moreover portrays the conventional methodologies which were utilized before, to manage information their impediments and how it is being overseen by the new approach Hadoop. It additionally portrays the working of Hadoop along with its pros cons and security on huge data.

Keywords Big data · Hadoop · MapReduce · SQL

1 Introduction

Big data is a trendy expression which speaks to the development of voluminous data of an association which surpasses the points of confinement for its stockpiling [1]. There is a need to keep up the huge information due to

- Increase of capacity limits
- Increase of preparing power
- Availability of information.

K. Kaur (✉)
Chandigarh Engineering College, Mohali, India
e-mail: kamalinder.cse@cgc.edu.in

V. Bharti
Chandigarh University, Ajitgarh, India
e-mail: inhod.ibmcse@cumail.in

© Springer Nature Singapore Pte Ltd. 2019
M. Mittal et al. (eds.), *Big Data Processing Using Spark in Cloud*,
Studies in Big Data 43, https://doi.org/10.1007/978-981-13-0550-4_1

1.1 Types of Data for Big Data

- Traditional enterprise data—incorporates data of client from CRM frameworks, value-based ERP information, web store exchanges and general record information [1].
- Machine-generated/sensor data—incorporates Call Detail Records ("CDR"), weblogs, keen meters, fabricating sensors, gear logs, exchanging frameworks information.
- Social information—incorporates client criticism streams posted by the general population over the world, small-scale blogging locales as Twitter, online networking stages like Facebook
- Stock Trading Data: The stock trade information holds data regarding 'purchase' and 'give' choices on common assets, offers of the clients are overseen from the organizations.
- Gridding Data: The power matrix information holds data devoured by a specific hub as for a base station
- Transport Data: Transport information incorporates show, limit, separation and accessibility of a vehicle.
- Search Web Engine Data: Search engines recover bunches of information from various tables [2].

1.2 Characteristics of Big Data

- Volume: Volume alludes to the measure of information. Machine-created information is delivered in significantly bigger amounts than non-conventional information [3]. For, e.g. a solitary stream motor can create 10 TB of information in 30 min. With more than 25,000 carrier flights for each day, the day-by-day volume of simply this single information source keeps running into the Petabytes. Savvy meters and substantial mechanical hardware like oil refineries and penetrating apparatuses create comparative information volumes, intensifying the issue.
- Velocity: Is the speed of preparing the information. The pace at which information streams in from sources, for example, cell phones, clickstreams, high-recurrence stock exchanging and machine-to-machine forms is huge and consistently quick moving [4].
- Variety: It alludes to the kind of data. Big information reaches out past organized information, for example, numbers, dates and strings to incorporate unstructured information, for example, content, video, sound, click streams, 3D information and log records [3].
- Value: The monetary estimation of various information differs fundamentally. Regularly there is great data covered up among a bigger assemblage of non-conventional information; the test is distinguishing what is profitable and after that changing and separating that information for examination [3] (Fig. 1).

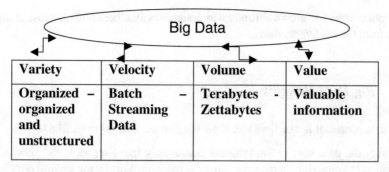

Fig. 1 V's of bigdata

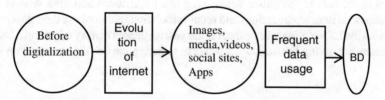

Fig. 2 Transformation from analogue to digital

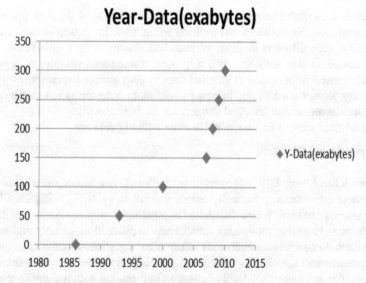

Fig. 3 Graph depicting the digital data usage since year 1980s

There is an expansion in data because we have shifted from p to digital [5]. There is a large data which is being shared on second basis with a person sitting in the corner of the world. This makes the hike of data usage and Internet packages for the purpose of making the data wide spread [4] (Figs. 2 and 3).

Data storage has grown abruptly which signifies that there is an increase in digital data from the analogous data.

1.3 Big Data Benefits

Big data is crucial in our lives and it gives a few advantages to us like [6]:

- Using the data kept in the informal community like Facebook, the showcasing offices are directing themselves about the reaction made by the general population for their battles, advancements, and other promoting mediums.
- Using the data in the online networking like inclinations and item view of their customers, item organizations and retail associations are booking their generation.
- Using the information with respect to the past medicinal history of patients, healing centers are giving the master frameworks to better and brisk administration [7].

1.4 Big Data Technologies

Big data is s an umbrella term portrays the accumulation of datasets that cannot be prepared utilizing conventional figuring techniques. In request to deal with the huge information different devices, systems and structures are required. Big information advancements are imperative to process tremendous volumes of organized and unstructured information in genuine time in giving more precise investigation, which may prompt a leadership bringing about more noteworthy transactional benefits, price decreases and lessened dangers for the business [3].

Two advance classes in the market dealing with big data are

(a) **Operational Big Data**

Framework like MongoDB: give operational abilities to continuous, intelligent workloads where information is basically caught and put away. NoSQL Big Data frameworks: intended to exploit new distributed computing structures stands over the previous decades to enable monstrous calculations to proceed reasonably and productively which do operational enormous information significantly simpler to oversee, less expensive and speedier to actualize [8]. NoSQL systems: provide experiences into examples and patterns in light of constant information with insignificant coding and without the need of information researchers and founders.

(b) **Analytical Big Data**

Massively Parallel Processing (MPP) database frameworks and MapReduce provides logical abilities for review and complex investigation. MapReduce gives another technique for dissecting information that is inverse to the capacities gave by SQL,

Table 1 Operational versus analytical systems

	Operational	Analytical
Time period	In miliseconds	In minutes
Concurrently	1000–100,000	1–10
Pattern access	Both read and write	Read operation
Queries	Selective	Unselective
Scope of data	Effective	Retrospective
Stock keepers	Customer	Data Scientist
Technology	Without SQL (NoSQL)	MapReduce, MPP record

Table 2 Stage 1 of hadoop

The java classes are as container record containing the execution of guide and lessen capacities
The area of the info and yield records in the appropriated document framework
The occupation setup by setting diverse parameters particular to the activity

and a framework in light of MapReduce may be scaled up by single servers to a large number of end machines [7] (Tables 1 and 2).

1.4.1 Basic Types of Big Data

Structured Data: Planned data are numbers and words that can be effectively classified and analysed. The information is produced by things like system sensors implanted in electronic gadgets, advanced cells and worldwide situating framework (GPS) gadgets. Structured data additionally incorporate things deals with figures, account adjusts and exchange information.

Unstructured Data: Unstructured Data incorporate more mind boggling data, for example, client audits from business sites, photographs and other mixed media, and remarks on long range informal communication locales. This type of information cannot be isolated and converted into classifications or dissected numerically. The touchy development of the Internet as of late implies that the assortment and measure of enormous information keep on growing. A lot of that development originates from unstructured data [9].

There are many challenges, out of which main are as follows:

- Creation
- Transfer
- Presentation
- Storing place
- Giving
- Capturing data
- Searching

- Comparison.

Organizations normally take the help of enterprise servers to fulfill the challenges.

2 Older Methodology

In this, a client would have a PC to process and keep huge information. In this information put away in Oracle Database, Sybase, Nutessa, MS SQL Server and DB2 and advanced virtual products may be composed with interface for the records, start processing the vital information and offer this the clients for examination reason [10].

Weakness
It functions admirably on adequate amount of information which may be upheld by common servers, or reaches up the computationer which handles the information. It is not reasonable to immense informational collections.

3 Hadoop

Google tackled the above-mentioned issue by utilizing a calculation called MapReduce. This calculation isolates the undertaking into little parts and out of those parts to numerous PCs associated with the system, and gathers the outcomes to shape the last outcome dataset. Doug Cutting and Mike Cafarella and in 2005 a group assign the arrangement gave by Google and started an Open Source called as HADOOP and Doug marked it hadoop after his child's toy elephant. Presently Hadoop is an enlisted benchmark of the Apache Software Foundation [5]. Hadoop utilizes the MapReduce calculation to run its applications where the information is handled in parallel on various CPU hubs. Hadoop is sufficiently competent to create applications which can keep working on bunches of desktops and they could perform finish factual investigation for a huge amount of information can be performed (Figs. 4 and 5).

Hadoop is java-based Apache open-source structure written in java. Hadoop is intended to move from single to many machines giving storage and calculation.

3.1 Hadoop Architecture

Structure of Hadoop has four modules which are as follows:

1. **Hadoop Common**: Hadoop Common utilities are the libraries in java language and are likewise utilized as utilities which help the other three modules of Hadoop.

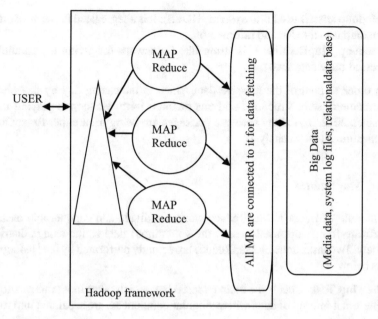

Fig. 4 Hadoop frame work

Fig. 5 Four modules in hadoop framework

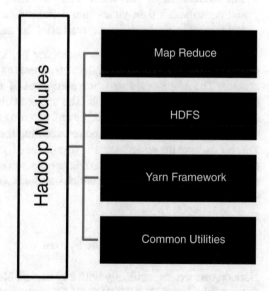

It contains the different fundamental scripting records keeping in mind the end goal to begin the working of Hadoop.

2. **Hadoop YARN**: It is the structure of Hadoop utilized for planning the occupations lying in pool and manage the assets which lies in cluster distribution.

3. **Hadoop Distributed File System (HDFS)**: It is a segregated framework offering an output to the demand information.
4. **Hadoop MapReduce**: It is string-like framework for giving the parallel processed huge data sets.

In order to control the massive data which is increasing day by day either in the structured, semi-structured and unstructured form, Hadoop is playing its best role since 2012. Too many software packages are being used either for streaming, optimization and data analysis.

3.1.1 MapReduce

The module of Hadoop, MapReduce works on divide-and-conquer approach as it is software framework used to split the voluminous data as the data is distributed parallely. Two tasks done by MapReduce are actually performed by the Hadoop itself are as below

- **The Map Task**: This is the main undertaking, which takes input information and believers it into set of data, where singular components are separated into rows of (key/value pairs).
- **The Reduce Task**: This errand takes the yield from a guide undertaking as info and consolidates data yields into a littler arrangement of tuples. The decrease assignment is constantly performed after the guide undertaking [11].

Regularly both the input and output are put away in a record framework. The system deals with booking assignments, checking them and re-executes the fizzled errands. The *MapReduce* structure comprises of a solitary ace *JobTracker* and one slave *TaskTracker* per group hub. The ace is in charge of asset administration, following asset utilization/accessibility and planning the occupations part undertakings on the slaves, observing them and re-executing the fizzled errands. The slaves *Task-Tracker* execute the assignments as coordinated by the ace and give errand status data to the ace intermittently. The *JobTracker* is a solitary purpose of disappointment for the Hadoop MapReduce benefit which implies if *JobTracker* goes down, every single running employment are ended.

3.1.2 Hadoop Distributed File System

Hadoop can work specifically with any mountable dispersed document framework, for example, Local FS, HFTP FS, S3 FS, and others, yet the most widely recognized record framework utilized by Hadoop is the Hadoop Distributed File System (HDFS) [12]. The Hadoop Distributed File System (HDFS) gives a circulated document framework that is intended to keep running on vast groups (a large number of PCs) of little PC machines in a solid, blame tolerant way. *HDFS* utilizes an ace/slave design where ace comprises of a solitary *NameNode* that deals with the document framework metadata and at least one slave *DataNodes* that store the genuine information. A document in a HDFS namespace is part into a few pieces and those squares are put away in an arrangement of *DataNodes*. The *NameNode* decides the mapping of squares to the *DataNodes*. The *DataNodes* deals with read and compose operation with the document framework. They likewise deal with piece creation, cancellation and replication in light of guideline given by *NameNode*. HDFS gives a shell like some other record framework and a rundown of charges are accessible to interface with the document framework. These shell summons will be shrouded in a different section alongside fitting cases [12–14].

3.2 Working of Hadoop

A client/application can present an occupation to the Hadoop (a Hadoop work customer) for required process by indicating the accompanying things:

Stage 2

The Hadoop work customer at that point presents the activity (jar/executable and so on) and arrangement to the *JobTracker* which at that point accepts the accountability of appropriating the product/setup to the slaves, planning assignments and observing them, giving status and demonstrative data to the activity customer.

Stage 3

The *TaskTrackers* use various hubs to execute the undertaking according to MapReduce usage and yield of the lessen work is put away into the yield records on the document framework.

3.2.1 Advantages of Hadoop

Hadoop enables the client to rapidly compose and test circulated frameworks. Though productive, programmed disperses the information and work over the pcs and thusly, uses basic distributed system of CPU centers

- It may not depend on equipments for giving adaptation to non-critical failure to accessibility (FTHA), rather at the application layer, library itself has intended to recognize and handles disappointment.
- Servers might be included or expelled out of group powerfully and it keeps on working with lack of intrusion.
- Another enormously preferred standpoint of Hadoop is that it is being separated from open source, it is good on every stages since it is Java-based.

3.2.2 Disadvantages of Hadoop

Hadoop has made the big data manageable. Due to its advantages still some draw-backs also prevail in the field of managing the data. As the coin has two sides, same is for the Hadoop. Along with some pros there are some cons of Hadoop too [15, 16]. Following are the disadvantages of Hadoop on which it is considered as a failure for them are as below:

a. Unable to manage small data files because of its high-speed capacity of processing in its design due to which it is unable to read and manage small files. This problem can be managed by simply clubbing the smaller files to make it as the bigger ones so that can be easily read by Hadoop's distributed file system. It can be also solved by adopting the sequence files by assigning the file name to the key value and the content of file as value.

b. Unable to manage the speed as MapReduce is doing the work in breaking into smaller parts which is time-consuming and later it is passed to reduce task where it is processed in less time but speed is slow in order to process it. To solve it Spark is used as it works on concept of in-memory reference processing as the data is processed in and out of disk directly. It is using a concept of spooling.

c. It does not support batch processing, rather Spark is used to do work in streaming.

d. It cannot be applied on real-time data. But again to solve it spark and flinkare are used.

e. It cannot be used for recursive processing of data. As cycles cannot be used to fetch the data while using Map () and reduce () functions of Hadoop. Hadoop can only work on single and small files [17, 18].

f. It is not easy to use as there is no iteration work done by Hadoop, whenever some iterative work is to be done then no option in Hadoop to use as we have to make the data set again with the new and separate values. There is no concept of backtracking due to non-iterative work flow in Hadoop. For this Hive and Pig can be used for it, otherwise the apache spark can be used [19, 20].

g. There is no concept of cryptography in Hadoop as it is lacking behind in securing the data. There is no security in data as no encryption takes place to encrypt the data. Sparks can be used to improve it.

h. There is no concept of data hiding and encapsulation in it. As each time a new activity is to be done and considered by removing the characteristics of one and adding it to another.

i. Hadoop follows the java platform as java is more prone to crimes and security breaches. Java, due to its independent platform in nature is more vulnerable to attacks.

j. Hadoop does not use the concept of caching the data as it directly stores in disk not in memory. It cannot cache the data in memory which makes it non-efficient tool for storing the big data [19].

k. The code of Hadoop is more vulnerable to bugs as its LOC is 1,20,000, which is time-consuming and not effective method to read the code and to remove the bugs from it.

All the drawbacks of Hadoop lead to advent usage of Apache Spark and Apache Flink and these are written in Scala and java, as Scala is the new replacement of java which came into existence in 2013. It is dealing the data in streamed form rather than that of batch form. It is also dealing with the streaming, machine learning and data analytics [21, 22].

3.3 Apache Spark

It is open-source software, can be used with the any big data tools. It can take the input data from Hadoop (MapReduce). It uses the batch process concept in collecting the data and can further do streaming on it. It can work in in-memory storage concept and can be further used on the clusters made by any big data tool. It works in streamline process as well as in iterative manner.

Apache spark works on DAG graphs data too. Due to its backtracking and cyclic approach it can work on directed acyclic graphs. It waits for the further instructions while giving the results of any problem. It is first solved using the cluster approach of Hadoop (batch processing) and further on moves on to stream the data and still waits for the final instructions to display the results which as a result apache spark saves the memory in short. It considers the scheduling the tasks itself whereas Hadoop requires external schedulers for scheduling it. Spark is fault tolerant [23].

3.3.1 Drawbacks of Apache Spark

   i. Due to its in-memory computation approach it has become costlier than any other big data tool.
  ii. It has no file management system as it relies on other big data tools.
 iii. It does optimization manually after applying the algorithm on it.

Due to the hindrances occurring in the path of Apache Spark there is a need to have emergence of some other platform which overcomes the problems of Hadoop as well as Spark. This gives rise to the emergence of Apache Flink [8].

3.4 Apache Flink

It is an optimized program; it gives result by applying optimization algorithm in the inbuilt application interface. It is faster than the spark as it does the processing of the part which actually gets changed. Flink returns the result with more latency and throughput as compared to spark and Hadoop. It is a pure streaming-based platform [18, 24]. It has its own memory management platform and it does not use the support of garbage collector of java. *Chandy–Lamport* distributed snapshot is used by Flink for the fault-tolerance mechanism. It is not using the *scalaplatform*. It supports checkpoints debugging tool feature which makes it a good recovery-based algorithm. Cost wise it is too expensive as compared to spark itself.

4 Security in Big Data

As information is expanding step by step there is a need to secure the voluminous information, however it is in structure, unstructured and semi-structure frame [9].

Security is typically an untimely idea, however elemental gives the correct innovation structure to the profound perceivability and multiple layers of security is being required on huge information venture. Multilevel assurance of information handling hubs implies actualizing security on the application, working framework while watching out for the whole framework utilizing noteworthy knowledge to stop any malignant action, developing dangers and vulnerabilities.

Key capabilities needed for secure dealing with data:

  i. Real-time connection and inconsistency identification of different security information
 ii. High-speed questioning of security knowledge information
iii. Flexible major information investigation crosswise over organized and unorganised information.
 iv. Visual application on instrument for envisioning and investigating big data
  v. Applications for profound perceivability.

4.1 Security Challenges on Big Data

  I. It improves Distributed Programming Framework by the securing the computations, in calculations and capacity to process enormous measures on information. A well-known case of delineate system, that parts info document into numerous lumps of primary period of guide lessen, a mapper for each piece peruses the information, play out some calculation, and yields a rundown of key/esteem sets. In the following stage, subtract or joins the qualities having a place with each particular key and yields the outcome [9].
 II. Security Practices on Non-Relational Data, Non-social information are promoted by No-SQL databases are as yet advancing concerning the security framework. For example, hearty answers for No-SQL infusion are as yet not developed; each NoSQL databases were worked to handle distinctive difficulties postured by the organizations. Designers utilizing NoSQL databases as a rule implant safety in the agents. Be that, grouping part of NoSQL databases represents extra tests for the heartbeat of such safety rehearses [25].

III. Transaction, Storage and Exchange backup files need security in layered forms to check data; physically moving content between levels gives control by supervisor coordinate. Meanwhile, the informational index has been measured and keeps on being, developing exponentially, adaptability and accessibility necessitude auto tiring for huge information stockpiling administration. Auto tiring arrangements do not monitor where the information is put away, which postures new difficulties to secure information stockpiling. New instruments are basic to approve access and look after all day, everyday availability. Business of such security rehearses.

IV. Enormous information utilize in entrepreneur settings require accumulation of content from different usages, for example gadgets, a safety data, administration framework might gather occasion backup files from a great many equipment gadgets and programming application in a venture arrange. Input approval is a key test in the information accumulation process [25].

V. Genuine—surveillance security tools observing has dependably been a test given the quantity of cautions produced by (security) devices. Various cautions (corresponded or not) tends to numerous pros and cons, for the most part overlooked or basically 'clicked away', as people cannot adapt to the shear sum. This issue may even increase with the offer information given the volume and speed of information streams notwithstanding, enormous information innovations may likewise give an open door, as in these advances do take into consideration quick preparing and investigation of various sorts of information. Which in its turn can be utilized to give, for example, ongoing irregularity location in light of adaptable security examination.

VI. Big analytics information may be viewed to an alarming indication leading to enormous sibling by conceivably empowering intrusions of protection, obtrusive advertising, diminished common opportunities and increment the control mechanism. Current examination of many companies is utilizing information investigation to advertise reason distinguished a case. Information to investigate is not sufficient to keep up client protection. For instance AOL-discharged anonymized scan logs for scholastic purposes yet clients were effectively recognized by the scholars [26, 27].

VII. To provide the assurance of the floating data is secure and just available for the approved substances, information must be encoded in light of access control strategies. To guarantee validation, assertion, decency among the appropriated elements, encryption secure structure must be executed.

VIII. Information Provenance metadata will develop many-sided quality because of extensive provenance diagrams created from provenance-empowered programming situations in huge information applications. Examination of such huge provenance diagrams to distinguish metadata conditions for security/classification applications is computationally serious.

5 Big Data Analysis Challenges

1. Incompleteness along with Heterogeneity: Some people devour data, a variation
    in data may occur. Truth be told, the subtlety and lavishness of characteristic
    dialect can give important profundity. Be that as it may, machine examination
    calculations expect homogeneous information, and cannot comprehend subtlety
    [28]. In result, information must be precisely organized as an initial phase in
    (or preceding) information examination. Now to illustrate, a sufferer who has
    different restorative methodology at a healing place, consider a record for each
    medicinal methodology or research place test, one dataset for the whole healing
    facility stay, or one data set for all whole duration of life clinic connections of the
    patient. Check if decisions have been progressively recorded and, on the other
    hand, progressively more noteworthy assortment. More prominent structure is
    probably going to be required by numerous (customary) information examination
    frameworks. Be that as it may, the less organized plan is probably going to be
    more viable for some reasons [27].
2. Scale: Obviously, the principal thing anybody considers is the size of big data
    [29]. Overseeing to furnish us with the assets expected to adapt to expanding
    volumes of information. Information volume is scaling speedier than process
    assets.
3. Opportuneness: The other consideration is speed. Outline of a framework is
    viably managing the measure is likely likewise to bring about a framework that
    can procedure a given size of informational index speedier. In any case, it is
    not recently that this speed is typically implied when one talks about velocity
    with regards to big data. The numerous circumstances in which the consequence
    of the investigation is mandatory. Consider if a deceitful charge exchange is
    considered, it ought to in a perfect world be hailed before the exchange is fin-
    ished—conceivably keeping the exchange from occurring by any means. Clearly,
    a full investigation of a client's buy history is not probably going to be attainable
    progressively. Or maybe, we have to create halfway outcomes ahead of time so
    a little measure of incremental calculation with new information can be utilized
    to touch base at a brisk assurance [30, 31].
4. Protection: The security of information related to big data. Appliances well-being
    records, some rule is representing to do and not to do. For other information,
    directions, especially in America, are minority compelling. Be that as it may,
    there is incredible open dread with respect to the wrong utilization of individual
    information, especially through connecting of information from different sources.
    Overseeing security is viably both a specialized and a sociological issue, which
    must be tended to mutually from the two viewpoints to understand the guarantee
    of enormous data [29].

6 Big Data Analytics—Security

It clarifies that big data is revolving the investigation scene. Specifically, big data investigation can be utilized to enhance data protection. For instance, big data examination can be utilized to break down money-related exchanges, log documents, and system movement to recognize inconsistencies and suspicious exercises, and to associate numerous wellsprings of data into an intelligible vie [32].

Information driven data security goes back to bank misrepresentation identification- and abnormality-based interruption location frameworks. Misrepresentation identification is a standout among the most unmistakable uses for big data examination. MasterCard organizations have led extortion discovery for a considerable length of time. Notwithstanding, the custom-constructed foundation to dig big data for misrepresentation discovery was not conservative to adjust for other extortion recognition employments. Off-the-rack big data devices and methods are dealing as thoughtfulness regarding examination for extortion location in social insurance, protection and different fields [32, 33].

With regards to information investigation for interruption identification, the accompanying advancement is expected

(a) First Generation: Intrusion recognition frameworks—Security modelers understood the requirement for layered security (e.g. receptive security and break reaction) in light of the fact that a framework with 100% defensive security is unthinkable.
(b) Second Generation: Security data and occasion administration (SIEM)—Managing cautions from various interruption location sensors and standards was a major test in big business settings. SIEM frameworks total and channel alerts from many sources and present significant data to security examiners.
(c) Generation III: Big data investigation in security—Big data devices can possibly give a huge progress in noteworthy security knowledge by diminishing the ideal opportunity for connecting, combining and contextualizing assorted security occasion data, and furthermore to correlate long haul recorded information for measurable uses [28].

Security in Networks

In an as of late distributed contextual analysis, Zions Bancorporation declared that it is utilizing Hadoop bunches and business knowledge instruments to parse a bigger number of information more rapidly than with conventional SIEM apparatuses. As far as they can tell, the amount of information and the recurrence examination of occasions are excessively for customary SIEMs to deal with alone. In their new Hadoop framework running questions with Hive, they get similar outcomes in around one moment [33].

Despite the difficulties, the gathering at HP Labs has effectively tended to a few big data investigation for safemoves, some are featured in the segment. Initial, an expansive scale chart deduction approach was acquainted with distinguish malware-tainted has in an endeavour organize and the noxious areas got to by the venture's hosts. In particular, a host-space get to diagram was developed from huge venture occasion informational indexes by including edges between each host in the endeavour and the areas went by the host. The chart was then seeded with negligible ground truth data from a boycott and a white rundown, and conviction engendering was utilized to appraise the probability that a host or space is pernicious. Examinations on a 2 billion HTTP ask for informational index gathered at an extensive endeavour, a 1 billion DNS ask for informational collection gathered at an ISP, and a 35 billion system interruption location framework ready informational index gathered from more than 900 ventures (that is, having constrained information named as ordinary occasions or assault occasions needed to prepare irregularity locators) [23, 31, 34].

DNS terabytes of occasions comprising of DNS billions are solicitations and reactions gathered at an ISP were dissected. The objective was to utilize the rich wellspring of DNS data to recognize botnets, pernicious spaces and different vindictive exercises in a system. In particular, includes that are demonstrative of perniciousness were distinguished. For instance, vindictive quick motion spaces tend to keep going for a brief timeframe, though great areas, for example, cm.edu last any longer and take steps to some topographically circulated Internet protocols. At that point, grouping methods (e.g. choice trees and bolster vector machines) were utilized to recognize tainted hosts and noxious spaces. The examination has officially recognized numerous noxious exercises from the ISP informational index [35].

Big data encryption and key administration undertakings trust—*Gemalto's kSafeNet* arrangement of information assurance arrangements let clients secure their enormous information organizations—whether its a Hadoop foundation, or a nonsocial (NoSQL) database, for example, *MongoDB or Couchbase* without hindering the examination instruments that make these arrangements vital. Moreover, *Gemalto*binds together these—and in addition a whole environment of accomplice encryption arrangements—behind an incorporated encryption key administration apparatus.

6.1 Hadoop Encryption Solutions

The *SafeNet* information assurance portfolio can secure information at numerous focuses in the Hadoop engineering—from Hive and *Hbase* to singular hubs in the information lake [23].

With *Gemalto,* clients have a decision. Fuse straightforward application-level security by means of APIs to ensure information without changing their database structure Pick a section-level answer for Hive that licenses ordinary questioning Pick a document framework-level arrangement with hearty strategy based access controls. Each Hadoop enormous information encryption and tokenization arrangement is completely straightforward to the end-client and configuration saving encryption usefulness implies that clients will keep on benefitting from the investigation instruments that draw additional incentive from developing information stores.

6.2 NoSQL Database Encryption Solutions

NoSQL databases are flexible database arrangements that are very much adjusted for extensive amounts of shifted information sorts. Since they include more than conventional database tables—utilizing objects and lists rather—they require an alternate way to deal with enormous information security. Clients would now be able to ensure information in any NoSQL database including driving database merchants, for example, *MongoDB, Cassandra, Couchbase and HBase.* Use document framework-level encryption answer for securing the records, envelopes and offers that contain the documents and questions listed in the NoSQL pattern [34, 36].

Combined with strategy-based access controls, clients hold a fine level of control in spite of the huge information volumes.

Application-level enormous information encryption or tokenization arrangements append security straightforwardly to the information before it ever is spared into the NoSQL diagram.

Operations stay straightforward to the end-client while the database holds its capacity to lead questions and convey information without diminishes in execution. The main enormous information security examination instrument merchants—Cybereason, Fortscale, Hexis Cyber Solutions, IBM, LogRhythm, RSA and Splunk—against the five fundamental variables basic for understanding the full advantages of these stages. As Hadoop is a broadly utilized huge information administration stage and related biological system, it is not amazing to see it utilized as the reason for various enormous information security investigation stages. Fortscale, for instance, utilizes the Cloudera Hadoop dispersion. This permits the Fortscale stage to scale directly as new hubs are added to the group.

IBM's QRadar utilizes a conveyed information administration framework that gives even scaling of information stockpiling. Sometimes, dispersed security data administration frameworks (SIEM) may just need access to neighborhood information, however in a few circumstances—particularly measurable examination—clients may need to look over the disseminated stage. IBM QRadar additionally consolidates an Internet searcher that permits looking crosswise over stages, and additionally locally. This huge information SIEM, in the interim, utilizes information hubs as opposed to capacity range systems, which limits cost and administration multifaceted nature. This appropriated stockpiling model in light of information hubs can scale to petabytes of capacity—those associations require huge volumes of long haul stockpiling. RSA Security Analytics additionally utilizes an appropriated, unified engineering to empower direct scaling. The expert work process in RSA's device tends to a basic need when scaling to vast volumes of information: organizing occasions and assignments to enhance the effectiveness of investigation. Hexis Cyber Solutions' Hawkeye Analytics Platform (Hawkeye AP) is based on an information stockroom stage for security occasion information. Notwithstanding having low level, versatile information administration—for example, the capacity to store huge volumes of information in records over different servers—it is pivotal to have instruments for questioning information in an organized way. Hawkeye AP is tuned to store information in a period apportioned manner that wipes out the requirement for internationally remaking files. It is likewise composed as a perused just database. This takes into account execution improvements, yet more vitally, it guarantees that information will not be messed with once it is composed. It is important that Hawkeye AP utilizes columnar information stockpiling—rather than push arranged capacity—which is improved for examination applications.

Support for various information sorts: Volume, speed and assortment are terms frequently used to portray huge information. The assortment of security occasion information represents various difficulties to information coordination to a major information security investigation item. RSA Security Analytics' answer is to utilize a particular design to empower the catch of different information sorts while keeping up the capacity to include different sources incrementally. The stage is intended to catch huge volumes of full system bundles, NetFlow information, endpoint information and logs. Now and then different information sorts suggest numerous security apparatuses. IBM's QRadar, for instance, has a defenselessness supervisor segment intended to coordinate information from an assortment of weakness scanners and enlarge that information with setting significant data about system use. IBM Security QRadar Incident Forensics is another strength module for investigating security episodes utilizing system stream information and full-bundle catch. The measurable device incorporates a web crawler that scales to terabytes of system information. LogRhythm's Security Intelligence Platform is another case of a major information security investigation stage with a far reaching support for various information sorts, including: framework logs, security occasions, review logs, machine information, application logs and stream information. The stage investigates crude information from these sources to create second-level information about record trustworthiness, process action, arrange correspondences, client and action. Splunk Enterprise Security enables examiners to look information and perform visual relationships to distinguish pernicious occasions and gather information about the setting of those occasions [37].

Versatile information ingestion: Huge information examination security items must ingest information frame servers, endpoints, systems and other framework segments that are continually evolving states. The important danger of this information ingestion segment is that it cannot stay aware of the inundation of approaching information. Enormous information security examination are equipped for breaking down a wide scope of information sorts while handling huge volumes of information. Splunk is generally perceived for its expansive information ingestion capacities. The stage offers connectors to information sources, as well as takes into account custom connectors also. Information is put away in a pattern less form and ordered on ingestion to empower differing information sorts while as yet giving quick inquiry reaction.

Another vital kind of reconciliation is information increase. This is the way toward adding logical data to occasion information as it is gathered. For instance, RSA Security Analytics improves arrange information as it is broke down by including insights about system sessions, danger markers and different points of interest that can enable investigators to comprehend the more extensive picture encompassing low-level security information [38].

Security investigation devices: Huge information security expository devices should scale to meet the measure of information created by an undertaking. Examiners, in the interim, ought to have the capacity to question occasion information at a level of reflection that considers the point of view of a data security outlook. Fortscale utilizes machine learning and factual investigation—all things considered known as information science systems—to adjust to changes in the security condition. These strategies enable Fortscale to drive investigation in light of information instead of just predefined rules. As gauge practices change on the system, machine learning calculations can identify the progressions without human mediation to refresh settled arrangements of principles. RSA Security Analytics incorporates predefined reports and standards to empower experts to rapidly begin making utilization of information gathered by the huge information investigation SIEM [39].

Security examination is likewise intensely subject to knowledge about malevolent exercises. RSA Security Analytics incorporates the RSA Live administration that conveys information preparing and connection guidelines to RSA Security Analytics organizations. These new standards can be utilized to break down new information landing progressively and recorded information put away on the RSA Security Analytics framework. Like Fortscale, RSA Security Analytics utilizes information science strategies to improve the nature of examination. LogRhythm's examination work process, in the mean time, incorporates handling, machine investigation and measurable examination stages. The handling step changes information in approaches to improve the probability that helpful examples will be distinguished from the crude information. This handling incorporates time standardization, information grouping, metadata labeling and chance contextualization. Consistence revealing, alarming and observing: Consistence detailing of some sort is an absolute necessity have prerequisite for most ventures today. Know that announcing administrations included with the enormous information security stages being considered by an association meet its particular consistence needs. IBM Security QRadar Risk Manager add-on gives apparatuses to oversee organize gadget setups in help of consistence and hazard administration. Capacities of the Risk Manger add-on include: computerized observing, bolster for various merchant item reviews, consistence arrangement appraisal and danger displaying [39].

7   Literature Review

Reena Singh, **KunverArif Ali**, 'In Challenges and Security Issues in Big Data Analysis' [Jan 2016] [22] the author describes about the various forms of data that can be used to describe the data which is increasing in number and Hadoop is used how it can overcome the problems faced by the user.

Shirudkar et al. International Journal of Advanced Research in Computer Science and Software Engineering 5(3), March—2015, pp. 1100–1109 [9] security methods are being implemented by the author and further he extended it to apply big data privacy on hybrid data through the data sets by encryption technology (type-based keyword search) and also described in it the malicious filtering of big data by including the lexical features.

Roger Schell "Security –'A Big Question for Big Data' in 2013 IEEE International Conference on Big Data https://doi.org/10.1109/bigdata.2013.6691547 Conference: 2013 IEEE International Conference on Big Data, he also explained about the prevailing methods that can help the user to provide security and privacy in the big data.

Harshawardhan S. **Bhosale**, **Prof**. **DevendraGadekar**, JSPM's Imperial College of Engineering and Research, Wagholi, Pune, (10–15 October, 2014), a review on big data and Hadoop, the paper has briefly described about the three v's of big data. The paper also focused on the problems faced by big data. It also describes the challenges and techniques applied on the application field of development which lay stress on the working and management of the Hadoop [3].

Shilpa, **Manjeet Kaur**, **LPU**, **Phagwara**, India, a review on Big Data and Methodology (5–10 October, 2013) deals with the challenges and issues regarding the big data. How the data is being accumulated, acquired and consolidated on the central cloud. It also describes how to mine the facts related to the data sets. It also encourage the user to read the data stored on the cloud which is too centralized and effectively solved [8].

Garlasu, **D**.; **Sandulescu**, **V**.; **Halcu**, **I**.; **Neculoiu**, G. (17–19 Jan. 2013), 'A Big Data implementation based on Grid Computing', Grid Computing served the favorable position dealing with capacity abilities and the Hadoop innovation is utilized for the execution reason. Matrix calculation gives the idea of appropriated figuring. The advantage of Grid registering focus is at hike stockpiling ability and preparing power is also at hike. Matrix Computing recognized the huge commitments among the logical research [2].

8 Conclusion

It is deduced that this part facilitates the extent of big data past the Hadoop can be effectively taken forward by flink and start. The security can be executed on the information to give the huge information, secure information to spread over the world. The measures are taken to give the security and which can be additionally extended further in future to give better information.

References

1. Roger Schell: Security—A Big Question for Big Data. In: 2013 IEEE International Conference on Big Data. https://doi.org/10.1109/bigdata.2013.6691547
2. Michael, K., Miller, K.W.: Big Data: New Opportunities and New Challenges. Published by the IEEE Computer Society 0018-9162/13/$31.00 © 2013 IEEE
3. Ji, C., Li, Y., Qiu, W., Awada, U., Li, K.: December. Big data processing in cloud computing environments. In: 2012 12th International Symposium on Pervasive Systems, Algorithms and Networks (ISPAN), pp. 17–23. IEEE (2012)
4. Muhtaroglu, F.C.P., Demir, S., Obali, M., Girgin, C.: Business model canvas perspective on big dataapplications. In: 2013 IEEE International Conference on Big Data, Silicon Valley, CA, 6–9 Oct 2013, pp. 32–37
5. Kaur, K., Kaur, I., Kaur, N., Tanisha, Gurmeen, Deepi: Big data management: characteristics, challenges and solutions. Int. J. Comput. Sci. Technol. JCST 7(4) (2016). ISSN 0976-8491, ISSN 2229-4333 (Print)
6. Tankard, C.: Big data security. Netw. Secur. 2012(7), 5–8 (2012)
7. Mahajan, P., Gaba, G., Chauhan, N.S.: Big data security. IITM J. Manag. IT 7(1), 89–94 (2016)
8. Bhosale, H.S., Gadekar, D.P.: A Review on Big Data and Hadoop, JSPM's Imperial College of Engineering & Research, Wagholi, Pune
9. Shirudkar et al.: A review on—big data: challenges, security & privacy issues. Int. J. Adv. Res. Comput. Sci. Softw. Eng. 5(3), 1100–1109 (2015)
10. Shirudkar, K., Motwani, D.: Big-data security, department of computer engineering VIT, Mumbai, India. Int. J. Adv. Res. Comput. Sci. Softw. Eng. 5(3) (2015)
11. Inukollu1, V.N., Arsi, S., Ravuri, S.R.: Security Issues Associated with Big Data in Cloud Computing. (IJNSA) 6(3), (2014)
12. Zhao, Y., Wu, J.: Dache: a data aware caching for big-data applications using the Map Reduce framework. In: 2013 Proceedings IEEE INFOCOM, Turin, 14–19 Apr 2013, pp. 35–39
13. Lu, J., Senellart, P., Lin, C., Du, X., Wang, S., Chen, X.: Optimal top-k generation of attribute combinations based on ranked lists. In: Proceedings of the 2012 International Conference on Management of Data. 2012, pp. 409–420
14. Aggarwal, C.C., Wang, H.: Managing and Mining Graph Data. Springer Publishing Company, Incorporated (2010)
15. Patel, A.B., Birla, M., Nair, U.: Addressing big data problem using hadoop and map reduce. In: 2012 Nirma University International Conference on Engineering (NUiCONE), 6 Dec 2012
16. Chen, J., Chen, Y., Du, X., Li, C., Lu, J., Zhao, S., Zhou, X.: Big data challenge: a data management perspective. In: Key Laboratory of Data Engineering and Knowledge Engineering. School of Information, Renmin University of China, Beijing, China
17. Ahmed, E.S.A., Saeed, R.A.: A survey of big data cloud computing security. Int. J. Comput. Sci. Softw. Eng. (IJCSSE), 3(1), 78–85 (2014)
18. http://www.thewindowsclub.com/what-is-big-data
19. Muhtaroglu, F.C.P., Demir, S., Obali, M., Girgin, C.: Business model canvas perspective on big dataapplications. In: Big Data, 2013 IEEE International Conference, Silicon Valley, CA, 6–9 Oct 2013, pp. 32–37
20. Big Data Working Group: Big Data Analytics for Security Intelligence, Sept 2013
21. Big Data Meets Big Data Analytics Three Key Technologies for Extracting Real-Time Business Value from the Big Data That Threatens to Overwhelm Traditional Computing Architectures (white paper)
22. Singh, R., Ali, K.A.: Challenges and security issues in big data analysis. Int. J. Innov. Res. Sci. Eng. Technol. 5, (1) (2016)
23. http://data-flair.training/blogs/apache-spark-tutorial/
24. https://en.wikipedia.org/wiki/Big_data
25. Security and privacy in the era of big data arenci/nationalconsortiumfordatascience white paper
26. http://www.sas.com/en_us/insights/big-data/what-is-big-data.html

27. http://www.sas.com/en_us/insights/big-data/what-is-big-data.html
28. Shilpa, M.K.: A Review on Big Data and Methodology. LPU, Phagwara, India
29. Labrinidis, A., Jagadish, H.V.: Challenges and opportunities with big data. Proc. VLDB Endowment **5**(12), 2032–2033 (2012)
30. https://en.wikipedia.org/wiki/Big_data
31. https://www.Big Data Analytics Tools Overview & Benefits _ Qubole.htm
32. https://safenet.gemalto.com/data-encryption/big-data-security-solutions/
33. http://www.darkreading.com/security-monitoring/167901086/security/news/232602339/a-case-study-in-security-big-data-analysis.html
34. Marko Grobelnikmarko.grobelnik@ijs.siJozef Stefan Institute Ljubljana, SloveniaStavanger, May 8th 2012
35. https://intellipaat.com/tutorial/spark-tutorial/
36. An Oracle White Paper June 2013 Oracle: Big Data for the Enterprise
37. http://www.darkreading.com/monitoring/a-case-study-in-security-big-dataanalys/232602339
38. Security issues associated with big data in cloud computing, (IJNSA), vol.6, no.3, may 2014))
39. http://searchsecurity.techtarget.com/feature/Comparing-the-top-big-data-security-analytics-tools

Big Data Streaming with Spark

Ankita Bansal, Roopal Jain and Kanika Modi

Abstract A stream is defined as continuously arriving unbounded data. Analytics of such real-time data has become an utmost necessity. This evolution required a technology capable of efficient computing of data distributed over several clusters. Current parallelized streaming systems lacked consistency, faced difficulty in combining historical data with streaming data, and handling slow nodes. These needs resulted in the birth of Apache Spark API that provides a framework which enables such scalable, error tolerant streaming with high throughput. This chapter introduces many concepts associated with Spark Streaming, including a discussion of supported operations. Finally, two other important platforms and their integration with Spark, namely Apache Kafka and Amazon Kinesis are explored.

Keywords Spark streaming · D-Streams · RDD · Operations · Structured streaming · Kafka · Kinesis · Pipeline · Integration with spark

1 Overview

Big data is defined for enormous complex data sets that require more sophisticated techniques apart from traditional data computing technologies. Big data and its challenges can be easily understood by five Vs: Volume, Velocity, Veracity, Variety, and Value. Big data is generated rapidly and is highly unstructured. This raw data is useless until it is converted into beneficial insights. There are two ways to process this

A. Bansal (✉)
Department of Information Technology, Netaji Subhash Institute of Technology,
New Delhi, India
e-mail: ankita.bansal06@gmail.com

R. Jain · K. Modi
Department of Computer Engineering, Netaji Subhash Institute of Technology, New Delhi, India
e-mail: roopal96@gmail.com

K. Modi
e-mail: 96kanu@gmail.com

© Springer Nature Singapore Pte Ltd. 2019
M. Mittal et al. (eds.), *Big Data Processing Using Spark in Cloud*,
Studies in Big Data 43, https://doi.org/10.1007/978-981-13-0550-4_2

data: Batch Processing and Streaming (or stream processing). In batch processing, the data collected over time is processed in batches, whereas processing is done in real time under stream processing.

Chapter Objectives

- To introduce streaming in Spark, its basic fundamentals
- Illustrating architecture of Spark
- To explore several types of operations supported on Spark API
- To describe various data sources where data can be ingested from
- To explore Kafka and Kinesis and their integration with spark
- To build a real-time streaming pipeline.

1.1 Streaming

A gigantic amount of data created by thousands of sensors and sending those data records simultaneously are defined as streaming data. This data requires processing on a record-by-record basis to draw valuable information. The analytics can be filtered, correlated, sampled, or aggregated. This analysis is in a form useful for various business and consumer aspects. For example, the industry can track most popular products amongst consumers based on corresponding comments and likes on social streams or can track sentiments based on some incidents, and take timely intuitive measures. Over the time, steam processing algorithms are applied to further refine the insights.

1.2 Real-Time Use Cases

Several interesting use cases include:

- Stock price movements are tracked in the real time to evaluate risks and portfolios are automatically balanced
- Improving content on websites through streaming, recording, computing, and enhancing data with users' preferences, to provide relevant recommendations and improved experiences
- Online gaming platforms collect real-time data about game-player intercommunication and analyze this data to provide dynamic actions and stimulus to increase engagement
- Streams to find the most trending topic/article on social media platforms, news websites
- Track recent modifications on Wikipedia
- View traffic streaming through networks

1.3 Why Apache Spark

Spark Streaming is becoming a popular choice for implementing data analytic solutions for the Internet of Things (IoT) sensors [1, 2]. Some of the benefits of using Apache Spark Streaming include:

- Improved usage of resources and balancing of load over traditional techniques
- It recovers quickly from stragglers and failures
- The streaming data can be combined with interactive queries and static data
- Spark is 100 times faster than MapReduce for batch processing
- Spark allows integration with advanced libraries like GraphX, Machine Learning, and SQL for distributed computing

Some of the most interesting use cases of Spark Streaming in the real life are:

1. Pinterest: Startup that provides tool for visual bookmarking uses Apache Kafka and Spark Streaming to gain insights about user engagement around the world
2. Uber: Employs Spark Streaming to collect terabytes of data from continuous streaming ETL pipeline of mobile users
3. Netflix: Uses Spark Streaming and Kafka to construct real-time data monitoring solution and online movie recommendation that processes billions of records each day from various sources
4. To find threats in real-time security intelligence operations
5. Triggers: To detect anomalous behavior in real time

1.4 Cloud-Based Apache Spark Platform

The increasing popularity of Apache Spark is leading to emergence of solutions built on cloud around Apache Spark:

- Hadoop based platforms provide support for Spark additionally to MapReduce [3, 4].
- Microsoft had added Spark support to its cloud-hosted version of Hadoop.
- Amazon Elastic Compute Cloud (EC2) can run Spark applications in Java, Python, and Scala [5].

2 Basic Concepts

This section covers fundamentals of Spark Streaming, its architecture, and various data sources from which data can be imported. Several transformation API methods available in Spark Streaming that are useful for computing data streams are described. These include operations similar to those available in Spark RDD API and Streaming Context. Various other operations like DataFrame, SQL, and machine learning

Fig. 1 Spark streaming

algorithms which can be applied on streaming data are also discussed. It also covers accumulators, broadcast variables, and checkpoints. This section concludes with a discussion of steps indispensable in a prototypical Spark Streaming program [6–9].

2.1 D-Streams

Earlier computational models for distributed streaming were less consistent, low level, and lacked fault recovery. A new programming model was introduced known as discretized streams (D-streams) that provide efficient fault recovery, better consistency, and high-level functional APIs [10–12]. Spark Streaming converts input live stream into batches which are later processed by Spark engine to produce output in batches. D-streams are a high-level abstraction provided by Spark Streaming. A D-Stream represents a continuous stream of data which supports a new recovery mechanism that enhances throughput over traditional duplication mitigates stragglers and parallel recovery of lost state.

Discretized streams are a sequence of partitioned datasets (RDDs) that are immutable and allow deterministic operations to produce new D-streams.

D-streams execute computations as a series of short, stateless, deterministic tasks. Across different tasks, states are represented as Resilient Distributed Datasets (RDDs) which are fault-tolerant data structures [13]. Streaming computations are done as a "series of deterministic batch computations on discrete time intervals".

The data received in every timestretch forms input dataset for that stretch and is stored in clusters. After the batch interval is completed, operations like map, reduce, and groupBy are applied on the dataset is processed to produce new datasets. The newly processed dataset can be a transitional state or a program outputs as shown in Fig. 1. These results are stored in RDDs that avoid replication and offer fast storage recovery (Fig. 2).

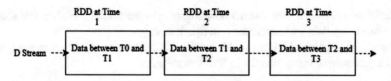

Fig. 2 Lineage graph for RDD

The following example will illustrate the idea of Spark Streaming. The program receives a stream from a server connected on a TCP socket and counts number of words in text data.

```
import org.apache.spark.streaming._
import org.apache.spark._
import org.apache.spark.storage._

# Create a local StreamingContext with two working thread and batch interval of 1 second
val sparkContext = new SparkConf().setMaster("local[2]").setAppName(" "WordCount")
batchIntervalSeconds = 1
streamingContext = StreamingContext(sc, batchIntervalSeconds)

val dStream = streamingContext.socketTextStream("localhost",8888)

val wordsDataset = dStream.flatMap(_.split(" "))
val pairDataset = wordsDataset.map(wordsDataset => (word, 1))
val wordCountsDataset = pairDataset.reduceByKey(_+_)

wordCountsDataset.print()

streamingContext.start()
streamingContext.awaitTermination()
```

2.1.1 Input D-Streams and Receivers

D-Streams that store stream of input data from streaming sources are known as Input D-Streams.An object that is processed in Spark's memory and stores data received from a streaming source is known as a Receiver.

Each input D-Streams is associated with a Receiver. For each data stream, a receiver is created and which runs on a worker node. Spark Streaming allows parallel processing of data streams by connecting to multiple data streams [4]. Receivers are further categorized as:

- Reliable receiver: It sends an acknowledgment to the sender (reliable data source) on receiving data and replicates it in Spark storage.
- Unreliable receiver: Unlike reliable receiver, it does not send any acknowledgment to a data source but replicates in Spark storage.

Selection of a receiver is done on the basis of chosen data sources and complexity of acknowledgment.

2.2 Data Sources

There are two kinds of built-in streaming sources in Spark Streaming:

- Basic Sources: StreamingContext API provides methods to create D-Streams from directly available sources. Only Spark Streaming library needs to be linked to the streaming application. These data sources are:
  - TCP Socket: Allows reading of data from a TCP source, hostname, and port need to be mentioned.
  - File Streams: Useful for reading files on file systems accordant to HDFS API (i.e., NFS, S3, HDFS, etc.). Files must not be altered once moved, as the newly appended data is not read.
  - Queue of RDDs: D-Streams can be created based on a queue of RDDs. Every RDD in the queue is processed and considered as a batch of data in D-Streams.
- Advanced Sources: Sources that are accessible via additional utility classes and require linking. These sources are not available in the shell. Corresponding dependencies are required to use them in Spark shell.
  - Some of the advanced sources are Apache Kafka [14], Amazon Kinesis [15], Twitter, Apache Flume, etc.
- Custom Resources: Implementation of the user-defined receiver is required.

2.3 Importing Data

In this section, an approach to obtain and export data in a production environment is discussed. Apache Spark was created to handle large volumes of data. There are two ways of batch importing data into Spark:

1. Batch Data Import: The entire dataset is loaded at once either from files or databases.
2. Streaming Data Import: Data is obtained continuously from streams.

2.3.1 Batch Import

In batch import, data is loaded from all the files at once. In production systems, Spark can be used to compute a batch job at fixed time intervals to process and publish statistics. Batch importing covers:

- Importing from files: The data should be available to all the machines on the cluster in order to facilitate batch import. Files only available on one worker node are not readable to others will cause failures. Following file systems are popular choices for massive datasets:

  - NFS or network file system assures that each of the machines can access the same files without replication. NFS is not fault-tolerant.
  - HDFS (Hadoop Distributed File System) gives every machine on the cluster access to files. The files are stored as a sequence of files and to make the system fault-tolerant, replication of blocks of a file is done.
  - S3 (Simple Storage Service) is an Amazon Web Services (AWS) solution to store large files in the cloud.

  Example: A sales record for some product is collected time to time by a company using S3 file system to store data. In a similar manner, data is collected for all the products. The company needs to know revenue generated each day. Thus the solution is to run Spark jobs daily on batch created by importing data for that particular day.

2.3.2 Streaming Import

Streaming has replaced the need for daily batch jobs that used to run on servers. There are two ways for stream importing in Spark:

- Built-In methods for Streaming Import: StreamingContext provides several built-in methods for importing data.

  - socketTextStream is used to read data from TCP sockets
  - textFileStream reads data from any Hadoop Compatible FileSystem Directory

- Kafka: In socket streaming, the files are copied after HTTP request. This ensures auto-refresh but is not perfectly real time. Kafka is a high efficiency distributed system that sends data immediately.
- Kinesis
- Twitter.

2.4   Supported Operations

This section covers various operations that are supported by streaming data. These include transformations, window operations, output operations, and lastly Spark DataFrame Spark SQL operations.

2.4.1   Transformation Operations on D-Streams

These operations allow modifications on the data in the input D-Streams. Although D-Streams functions RDD methods have same name but they are different (Table 1).

2.4.2   Window Operations

Spark Streaming allows transformations to be applied in sliding window manner. These computations are known as windowed computations. Window Operations

Table 1 Some of the transformation operations allowed on D-streamss

Function	Utility
map(function)	Returns a new D-Streams by mapping each source element
flatmap(function)	Each input entry can be mapped to multiple output entries
filter (function):	Selects records from source D-Streams on basis of function and returns a new D-Streams
repartition(numberOfPartitions)	change number of partitions and control level of parallelism
union(diffStream)	Combines source D-Streams with diffStream and returns a new D-Streams
count()	Counts number of entries in each RDD and returns a new D-Streams of RDDs containing a single element
reduce(function)	Using a commutative and associative functio, a new D-Streams is returned which contains aggregated elements of each source D-Streams RDD
countByValue()	A new D-Streams is returned in which each entry represents key and its frequency
reduceByKey(function, [numberOfTasks])	A new D-Streams is returned in which the given reduce function is used to evaluate aggregate values for each key
join(diffStream, [numberOfTasks])	A new D-Streams is returned with every pair for each key
transform(function)	This method is used to apply a RDD-to-RDD operations to each RDD in the source D-Streams
updateStateByKey(function)	This method is useful for maintaining arbitrary state data for all keys. A new D-Streams is returned where the previous state of each key is updated using given function

are: useful to find out what has happened in some given time frame for example in the last one hour and those statistics are required to be refreshed every minute. The source RDDs falling within the window are merged to form RDD of that particular window which is used for computation purposes.

All window operations take *slideInterval(sI)* and *windowLength(wL)* as parameters (Table 2).

One interesting application of window operation can be to generate word counts every 5 s over the last 40 s (Table 3).

Table 2 Common window operations

Transformation	Meaning
window(wL, sI)	A new D-Streams is returned containing windowed batched of original D-Streams
countByWindow(wL, sI)	Sliding window count of elements is returned
reduceByWindow(function, wL, sI)	Using a commutative and associative function that can be computed in parallel, a new D-Streams is returned which contains aggregated elements of in-stream over sliding window
reduceByKeyAndWindow(function, wL, sI, [numberofTasks])	A new D-Streams is returned in which the given reduce function is used to evaluate aggregate values over batches in the sliding window
reduceByKeyAndWindow(function, inverseFunc, wL, sI, [numberofTasks])	This method is enhanced version of reduceByKeyAndWindow() which calculates the reduce value of each window incrementally, making use of reducing values of preceding window
countByValueAndWindow(wL, sI, [numberOfTasks])	A new D-Streams is returned in which each entry represents key and its frequency in the sliding window

Table 3 Output operations available in spark

OutPut Operation	Utility
print()	Useful for debugging and development. First ten elements of each batch of D-Streams are printed on the driver node which is used for running the application
saveAsTextFiles(prefix,[suffix])	D-Streams's content is saved as text files. At each batch interval, on the basis of prefix and suffix, file name is generated
saveAsObjectFiles(prefix,[suffix])	D-Streams's content is saved as SequenceFiles of serialized Java objects. At each batch interval, on the basis of prefix and suffix, file name is generated
saveAsHadoopFiles(prefix,[suffix])	D-Streams's content is saved as Hadoop files. At each batch interval, on the basis of prefix and suffix, file name is generated
foreachRDD(function)	This function is used to push the data to an external system available in each RDD. External systems can be writing over the network to database or saving the RDD to files

2.4.3   Join Operations

Spark Streaming allows joining operations between D-Streams. Allowed four join operations are: join, leftOuterJoin, rightOuterJoin, fullOuterJoin
Stream-Stream Join

```
val s1: DStream[String, String] = .....
val s2: DStream[String, String] = .....
val streamJoin = s1.join(s2)
```

The RDD generated by stream2 will be combined with the RDD generated by stream1 in each batch interval. Spark Streaming also allows performing joins over windows of streams.

```
val windowStream1 = s1.window(60)
val windowStream2 = s2.window(80)
val windowJoin = windowStream1.join(windowStream2)
```

Stream-dataset Join

```
val dataset = ... # Some RDD
val stream = ... # Some Stream

val finalDStream= stream.transform { dataset => dataset.join(dataset) }
```

Windowed Streams can be joined in a similar manner. Datasets to be joined can be dynamically changed as for every batch interval, function provided to transform is evaluated.

```
val windowstream = stream.window(30)
val joinedDstream = windowStream.transform{ dataset => da-
taset.join(dataset) }
```

2.4.4   Output Operations

D-Stream's data can be consumed by external systems like file systems or databases. Output operations trigger the execution of all D-Streams transformations as they allow the transformed data to be pushed out to external systems. Just like RDDs, D-Streams are computed lazily by the output operations. Output operations are executed one-at-a-time by default and in the same order they are defined in the program. Following output operations are available in Spark:

SaveAsObjectFiles() and SaveAsHadoopFiles() are not available in Python API.

2.4.5 SQL and DataFrame Operations

Spark SQL and DataFrame operations can be easily applied on streaming data. This requires creating a SparkSession using the SparkContext used by the StreamingContext. A lazily instantiated singleton instance of SparkSession is created to ensure restarting on driver failures. Every RDD is converted to a DataFrame for application of SQL queries.

```
val user_rdd = DStream[String]   #RDD containing some user information
using a website
val user_df = rdd.toDF("user_rdd")
val count_df = spark.sql("select user, count(*) as total from user group by
user")
count_df.show()
```

SQL queries can also be run on tables defined on streaming data asynchronously to the running StreamingContext. Batch duration of the StreamingContext should be set enough to remember data for query to be run. Otherwise, some data gets deleted before running query is complete.

2.4.6 Machine Learning Operations

Machine learning Library (MLlib) provides a large class of the machine learning algorithms that can be trained using historical data and can be applied on streaming data online. There are few streaming machine learning algorithms that are capable of training simultaneously from the streaming data and test the model on streaming data. One such streaming algorithm is Streaming KMeans which updates the cluster dynamically as the new data arrives. Another streaming algorithm is Streaming Linear Regression which updates the parameters in the real time. The following example shows the usage of Streaming Linear Regression on streaming data.

```
val totalFeatures = 3
val model = new StreamingLinearRegressionWithSGD()
  .setInitialWeights(Vectors.zeros(totalFeatures))

model.trainOn(data_train)
model.predictOnValues(data_test.map(lp => (lp.label, lp.features))).print()

streamingContext.start()
streamingContext.awaitTermination()
```

```
import
org.apache.spark.mllib.regression.StreamingLinearRegressionWithSGD
import org.apache.spark.mllib.regression.LabeledPoint
import org.apache.spark.mllib.linalg.Vectors

val sparkContext = new SparkConf().setMaster("local[2]").setAppName("
"LinearRegression")

val data_train= ssc.textFileStream(args(0)).map(LabeledPoint.parse).cache()
val data_test = ssc.textFileStream(args(1)).map(LabeledPoint.parse)
```

The program reads input data from a text file and model is initialised by setting the weights to zero. Total number of features is three.

2.5 Steps in Spark Streaming Program

A classic Spark Streaming program involves following steps:

- Initialize StreamingContext object with the parameters SparkContext and sliding interval time. The sliding interval is used to set update window. Once initialization is done, new computations cannot be defined to already existing context.
- Next, input D-Stream is created for specifying input data sources.
- After input D-Streams is created, computations are defined using Spark Streaming transformation APIs.
- Once streaming computation logic is defined, processing starts using *start* method in StreamingContext
- Finally, processing is terminated using StreamingContext method *awaitTermination.*

3 Real-Time Streaming

3.1 Structured Streaming

As the streaming system became more complex, Apache Spark came up with the idea of Structured Streaming in Spark version 2.0. Structured streaming is not a huge change to Spark itself but a collection of additions. The fundamental concepts remain same; the core transformation is from RDD to DataFrames as before. In this section, aspects of Structured Streaming are discussed.

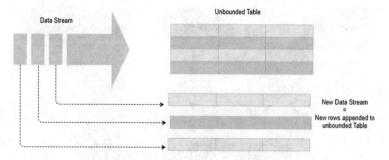

Fig. 3 Fundamentals of structured streaming

3.1.1 Basic Concepts

Structured Streaming introduces the concept of *unbounded table*. Consider the input stream as an Input table. As the data keeps on arriving, new rows are being appended to the Input table (Fig. 3).

In Structured Streaming, DataFrames and Datasets represent streaming, unbounded data as well as static, bounded data. Streaming DataFrames/Datasets are created using the same entry point for static DataFrames/Datasets and can be transformed or operated in exact same manner as batch DataFrames using user-defined functions (UDFs). Structured streaming provides unified API for both streaming and batch sources.

The new model can be explained by the following terms:

- Input: Data from source as an append-only table
- Trigger: Frequency of checking input for new data
- Query: Operations on input like map/reduce/filter/window/session operations
- Result: Final operated table is updated every trigger interval
- Output: Which part of output to be written to data sink after every trigger

  - Complete output: Result table as an output is written every time
  - Delta output: Only modified rows are written
  - Append Output: Only new rows are appended.

3.1.2 Supported Operations

Almost all operations like RDD operations, SQL operations, etc., can be applied on streaming DataFrames/Datasets. Some of the basic operations are

- Aggregation, Projection, Selection
- Window Operations
- Join Operations
- Arbitrary Stateful operations like mapGroupWithState. These operations allow user-defined code to be applied to update user-defined state.

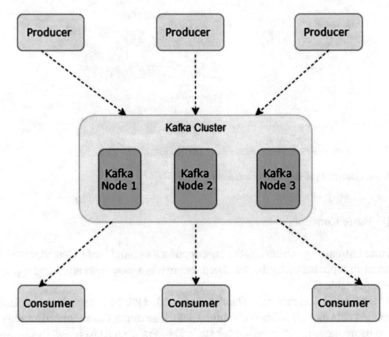

Fig. 4 Kafka

- Handling Late data

  However, there are some operations that Streaming DataFrames/Datasets do not support some operations. Few of them are:

- Chain of aggregations on a streaming DataFrame
- Distinct Operations
- Sorting operations are only supported after aggregation in Complete Output Mode.
- Any join between two streaming datasets is not yet supported.
- Outer joins between a static and streaming Datasets are conditionally supported
- Some Dataset methods that run queries immediately and return result like count(), show(), foreach() do not work on streaming Datasets.

3.2   Apache Kafka

Apache Kafka is a scalable, fault-tolerant, real-time messaging system for transferring data in real time between server, applications, and processes. It is used in the systems where capturing user activity on stock ticker data, logs [2], and websites is required. In this section, brief overview of Kafka is given and later on, its integration with Spark is discussed (Fig. 4).

3.2.1   Overview

Kafka is a software where applications connected to the system can transfer and process messages onto a category/feed name known as topic. Byte arrays that are capable of storing any object in any format are known as messages. A message is a stream of records. Each Kafka message is organized into topics. Each record has a key, a value, and a timestamp. Kafka works like a distributed database. When a message is posted to Kafka, it is written to the disk as well as replicated to different machines in the cluster simultaneously.

Kafka Broker: One or multiple servers known as Kafka brokers run a Kafka cluster.
  Kafka is based on four core APIs

1. Producer API: allows an application to push messages (publish data) to one or more topics within the broker.
2. Consumer API: allows an application to pull messages from one or more Kafka topics and process the stream of records.
3. Stream API: allows an application to take input stream from one or multiple topics and creating an output stream to one or more topics. In shorter words, this API converts input streams to output streams
4. Connector API: an API to build and run of reusable consumers or producers that connect Kafka topic to current data systems or applications.

Apache Kafka provides a data transfer framework known as Kafka Connect apart from client APIs.

Kafka Connect: Apache Kafka provides a framework for streaming data between external data systems and Apache Kafka to support data pipelines in organizations. Kafka Connect can be run as a distributed service or as a standalone process. In distributed service, it uses REST API to submit the connectors to Kafka Connect cluster.

3.2.2   Kafka Topic Partition

Kafka topics are divided into partitions that contain messages in an unmodifiable sequence. In a partition, each message is assigned a unique offset which is used for its identification as well. For multiple consumers to read from a single topic in parallel, it needs to have multiple partition logs where each partition must have one or more replicas. Messages with the same key arrive at same partition. Moreover, a topic can be parallelized by splitting the data across several servers.

Consumers can read data in any order. This feature of Kafka implies that consumers can come and go without little impact on other consumers and the cluster. Every partition is replicated across a number of servers to ensure fault tolerance.

3.2.3 Kafka as a Storage System

Kafka writes data to disk and duplicates it for increasing fault tolerance. Write operation is not complete unless replication is done and producers are made to wait for an acknowledgment. Kafka allows consumers to control their read position, i.e., moving backward or forward in a partition and takes storage seriously.

3.2.4 Kafka for Streaming Processing

Apart from reading, writing, and storing streams of data, Kafka can be used for real-time processing of streams. A stream processor takes continuous streams of data from input topics and produces continuous streams to output topics.

Simple stream processing can be done through consumer and producer APIs. Nonetheless, for complex processing operations like joining streams or aggregating streams, Kafka provides integrated Stream APIs as discussed in Sect. 3.2.1.

3.2.5 Advantages of Kafka

- Every topic can do scale processing and multi subscribing simultaneously as there is no trade-off between the two.
- It ensures ordering as messages in a particular topic partition are appended in the order they are send by the producer. Message sent earlier has a lower offset.
- In traditional queues, ordering is lost as records are consumed asynchronously. On the other hand, traditional messaging systems that allow only one process to consume imply no parallel processing can take place. But Kafka ensures ordering as well as load balancing. Parallelism takes place through partition within the topics. If there are N partitions, consumer group can consume a topic with a maximum of N threads in parallel. This is the maximum degree of consumer parallelism.
- By combining low-latency and storage capabilities, Kafka is suitable for building streaming data pipelines as well as streaming applications.

3.3 Integrating Kafka with Spark Streaming

Apache Spark Streaming applications that read messages from one or more Kafka topics have several benefits. These applications gather insights very fast, lead to uniform management of Spark cluster, guarantee delivery without the need of a write-ahead log and have access to message metadata.

3.3.1 Reading from Kafka

There are two ways to determine read parallelism for Kafka

1. Number of Input D-Streams: Spark runs one task (receiver) per input D-Streams; multiple D-Streamss will parallelize read operations across multiple cores/machines.
2. Number of consumer threads per input D-Streams: Same receiver (task) runs multiple threads, thereby parallelising read operations on the same machine/core.

Option 1 is preferred in practical usage as read-throughput is not increased by running several threads on the same machine.

3.3.2 Message Processing in Spark

After receiving data from Kafka, the data needs to be processed in parallel. More threads for processing can be used rather than reading. In Spark, parallelism is defined by number of RDD partitions as one receiver is run for each RDD partition. There are two ways to control parallelism here:

1. Number of Input D-Streams: Either they can be modified or taken as it is as they are received from previous stage.
2. By applying repartition transformation: repartition method can be used to change number of partitions, hence level of parallelism.

3.3.3 Writing to Kafka

foreachRDD operator is the most generic output operator and should be used for writing to Kafka. This function should be used to push data in RDD to an external system writing it over the network or saving into a file. New Kafka producers should not be created for each partition as Spark Streaming creates several RDDs containing multiple partitions each minute.

3.3.4 Apache Kafka on AWS

Kafka application can be run on Amazon Elastic Computing Cloud (EC2) offering high scalability and enhanced performance solutions for ingesting streaming data (Fig. 5).

Send streaming data to   Kafka buffers and saves   Use Spark Streaming to   Analytics available to
    Kafka topics   for consumers   transform, filter or   end users through
                                                                       aggregate   dashboard

Fig. 5 Apache Kafka on AWS

3.3.5 Real-Time Spark Streaming Pipeline Using Kafka

The example discussed below reads from a Kafka topic using Twitter. Number of tweets per user is counted within each batch. Later, same is computed over a time interval. Step by step program is explained [16].

Import necessary pySpark modules

```
#Spark Streaming
from pyspark.streaming import StreamingContext
#Kafka
from pyspark.streaming.kafka import KafkaUtils
#Spark
from pyspark import SparkContext
#json parsing
import json
```

Create Spark Context which is the starting point of a Spark Streaming program. Create StreamingContext. SparkContext is passed with batch duration of 60 s.

```
sparkContext = SparkContext(appName="KafkaSparkStreaming")
streamingContext = StreamingContext(sparkContext, 60)
```

Connection to Kafka cluster is made using native Spark Streaming capabilities. Consumer group is *spark_streaming* and topic connected is *twitter*

```
kafkaStream = KafkaUtils.createStream(streamingContext, 'cdh57-01-node-01.moffatt.me:2181', 'spark_streaming', {'twitter':1})
```

Input Stream is a D-Streams which is used to parse input messages stored in JSON format.

```
inputDS = kafkaStream.map(lambda v: json.loads(v[1]))
```

Tweets are stored in a JSON structure. Analysis is done by author name which can be accessed through *user.screen_name*

```
#To count and print number of tweets in a batch
#pprint prints 10 values and nothing is printed until start() is called

inputDS.count().map(lambda x:'Number of tweets in the batch are: %s' %
x).pprint()

user_stream = inputDS.map (lambda tweet: tweet['user']['screen_name'])

#Count the number of tweets done by a user
user_counts = user_stream.countByValue()
user_counts.pprint()
```

Sort the author on the basis of number of Tweets. D-Streams does not have an inbuilt sort function. So, *transform* function is used to access sortBy available in pyspark.

```
sorted_user_count = user_counts.transform(\
     (lambda x :x\
              .sortBy(lambda x:( -x[1]))))

sorted_user_count.pprint()
```

To get top five users with tweet count

```
top_five_users = sorted_user_count.transform\
     (lambda rdd:sc.parallelize(rdd.take(5)))

top_five_users.pprint()
```

To get authors whose username starts with 'r' and who have tweeted more than once

```
query1 = user_counts.filter(lambda x: x[1]>1
                    or
                    x[0].lower().startswith('r'))
```

To get most commonly used words in tweets

```
inputDS.flatMap(lambda tweet:tweet['text'].split(" "))\
  .countByValue()\
  .transform(lambda rdd:rdd.sortBy(lambda x:-x[1]))\
  .pprint()
```

Starting the streaming context: The program starts here and results of pprint() are printed.

```
streamingContext.start()
streamingContext.awaitTermination(timeout = 180 )
```

The complete program is given below

```
from pyspark.streaming import StreamingContext
from pyspark.streaming.kafka import KafkaUtils
from pyspark import SparkContext
import json

sparkContext = SparkContext(appName="KafkaSparkStreaming")

streamingContext = StreamingContext(sparkContext, 60)

kafkaStream = KafkaUtils.createStream(streamingContext, 'cdh57-01-node-
01.moffatt.me:2181', 'spark_streaming', {'twitter':1})
inputDS = kafkaStream.map(lambda v: json.loads(v[1]))

inputDS.count().map(lambda x:'Number of tweets in the batch are: %s' %
x).pprint()

user_stream = inputDS.map (lambda tweet: tweet['user']['screen_name'])

user_counts = user_stream.countByValue()
user_counts.pprint()
sorted_user_count = user_counts.transform(\
        (lambda x :x\
                .sortBy(lambda x:( -x[1]))))

sorted_user_count.pprint()

top_five_users = sorted_user_count.transform\
(lambda rdd:sc.parallelize(rdd.take(5)))
```

For Windowed Stream Processing, complete program looks like

```
top_five_users.pprint()

query1 = user_counts.filter(lambda x: x[1]>1
                or
                x[0].lower().startswith('r'))

inputDS.flatMap(lambda tweet:tweet['text'].split(" "))\
  .countByValue()\
  .transform(lambda rdd:rdd.sortBy(lambda x:-x[1]))\
  .pprint()

streamingContext.start()
streamingContext.awaitTermination(timeout = 180 )
from pyspark.streaming import StreamingContext
from pyspark.streaming.kafka import KafkaUtils
from pyspark import SparkContext
import json

sparkContext                        =                        SparkCon-
text(appName="KafkaSparkStreamingWindowed")

streamingContext = StreamingContext(sparkContext, 5)

kafkaStream = KafkaUtils.createStream(streamingContext, 'cdh57-01-node-
01.moffatt.me:2181', 'spark_streaming', {'twitter':1})
inputDS = kafkaStream.map(lambda v: json.loads(v[1]))

#Number of tweets in batch
batch_count = kafkaStream.count().map(lambda x:'Number of tweets in the
batch are: %s' % x).pprint()

#number of tweets in batch
window_count     =     kafkaStream.countByWindow(60,5).map(lambda
x:('Tweets total (One minute rolling count): %s' % x))

#Get users
user_stream = inputDS.map (lambda tweet: tweet['user']['screen_name'])
#word count in batch
batch_occurence = user_stream.countByValue()\
                .transform(lambda rdd:rdd\
                .sortBy(lambda x:-x[1]))\
                    .map(lambda  x:"User  counts  this  batch:\tValue
%s\tCount %s" % (x[0],x[1]))
```

```
window_occurence = user_stream.countByValueAndWindow(60,5)\
                .transform(lambda rdd:rdd\
                .sortBy(lambda x:-x[1]))\
                        .map(lambda x:"user counts in one minute win-
dow:\tValue %s\tCount %s" %

streamingContext.start()
streamingContext.awaitTermination(timeout = 180 )
```

3.4 Amazon Kinensis

Amazon Kinesis [15] Streams makes real-time streaming data simpler to collect, process, and analyze, gain timely insights and make quick decisions. Kinesis makes streaming cost effective, scalable and gives control to choose tools in the best interest of the application (Fig. 6).

A classical Amazon Kinesis Streams application takes input data from Kinesis streams as data records. Processed records can be used for a variety of applications like sending data to other AWS services, advertising strategies, dynamically change pricing or generating alerts and is sent to dashboards.

3.4.1 Benefits of Using Kinesis

- Real time: Processing and analyzing of data in real-time instead of waiting to have all the data collected.
- Fully Managed: It does not require the user to manage any infrastructure or any custom code.
- Scalable: Allows changing capacity of streams by API calls or by few clicks

Fig. 6 Using Amazon Kinesis on spark

- Reliable: Manual configuration for replication is not required as it automatically replicates across three AWS zones.
- Integration with open-source frameworks like Apache Flink and Apache Spark and with other AWS services makes loading and processing of data easier.

3.4.2 Kinesis Use Cases

Kinesis streams are used for continuous, rapid intake and aggregation of data. The processing is lightweight as the processing and response time for data ingestion is being done in the real time. Kinesis streams offer quick data input as data is not divided into batches on the servers before submitting for input. The typical use cases of Kinesis streams are:

- Kinesis can be used for complex stream processing as Directed Acyclic Graphs (DAGs) can be created from Kinesis Stream data streams and applications.
- Parallel processing on real-time data can be used for obtaining real-time analytics. For example, analyzing site usability by processing website clickstreams in real time.
- Accelerated data intake and processing: Data pushed directly into a stream can be processed in merely seconds. This ensures logs are not lost.
- Kinesis Streams are used for obtaining real-time metrics which are used for reporting the status of system in real time.

3.5 Kinesis on Spark

Kinesis can be used as a source for streaming data. It is not necessary to use Kinesis Analytics; Spark SQL and structured APIs can also be used instead. Processing Kinesis Streams with Spark Streaming requires an Elastic MapReduce (EMR) cluster with Spark to read data from Kinesis stream.

3.5.1 Create Kinesis Data Stream

Amazon Kinesis Client Library (KCL) is used for creating an input D-Streams via Kinesis consumer.

The following command is run from command line to create a Kinesis stream in user's account.

```
aws kinesis create-stream –stream-name my_stream –shard-count1
```

3.5.2 Kinesis Spark Streaming Application

A simple streaming application is discussed below which reads a Kinesis stream that contains events in JSON format. Spark job counts the events on basis of type and these counts are aggregated into 1 min buckets.

```
import org.apache.spark.streaming.kinesis.KinesisUtils
import org.apache.spark.SparkConf
import org.apaches.spark.streaming._

val streamingSparkContext = {
    val sparkConf = new SparkConf().setAppName( config.appName )
    .setMaster(config.master)
    New StreamingContext(sparkConf, config.batchInterval)
    }
    streamingSparkContext
```

Reading from Source: reading stream using Kinesis connector.

```
val   kinesisClient   =   KinesisUtils.setupKinesisClientConnection(  con-
fig.endpointURL, config.awsProfile)
require(kinesisClient != null, "No AWS credentials found")

val sreamingSparkContext = setupSparkContext(config)
val   numberShards   =   KinesisUtils.getShardCount(kinesisClient,   con-
fig.streamName)
val sparkStream = ( 0 until numberShards).map { i=>
    KinesisUtils.createStream(
            sc = streamingSparkContext,
            streamName = config.streamName,
            endPointURL = config.endpointUrl,
            initialPosInStream = config.initialPosition,
            checkPointInterval = config.batchInterval,
            storageLevel = config.storageLevel
            )
    }
```

Exploring and applying Transformation on Streams: After Kinesis records are mapped to DataFrames and data is loaded, Spark DataFrame and SQL APIs can be applied to obtain analytics. This is the part where insights are derived.

```
//Map phase:
val bucketEvent = streamingSparkContext.union( sparkStream)
            .map { bytes =>
                    Val e = SimpleEvent.fromJson(bytes)
                    (e.bucket, e.type)
}

//Reduce Phase
val tot_count  = bucketEvent.groupByKey
    .map { case (eventType, events) =>
    val count = events.groupBy(identity).mapValues(_.size)
    (eventType, count)
    }
```

Saving transformed Stream: Optionally, transformed data can be written some-where. In this example, finally aggregation is done and saved into Amazon DynamoDB

```
// add import for DynaoDB
import com.amazonaws.services.dynamodbv2.document.DynamoDB
import storge.DynamoUtils

//setup Spark Connection with DynamoDB
val conn = DynamoUtils.setupDynamoClientConnection(config.awsProfile)

//aggregating and storing
tot_count.foreachRDD{ rdd=>
    rdd.foreach { case (df, aggregate) =>
            aggregates.foreach { case (type, count) =>
                    DynamoUtils.setOrUpdateCount(
                            conn,
                            config.tableName,
                            df.toString,
                            type,
                            DynamoUtils.timeNow()
                            DynamoUtils.timeNow(),
                            count.toInt
                    )
            }
    }
}
```

3.6   Apache Kafka versus Amazon Kinesis

There are number of options available to work with streaming data on cloud. If cloud being used is AWS, then using Amazon Kinesis offers several advantages over Kafka. Both Amazon Kinesis and Apache Kafka are scalable, reliable and durable data ingesting framework sharing common core concepts like application components, partitioning and replication.

However, the following points should be considered before building a real-time application.

3.6.1   Configuration and Setup

Amazon Kinesis is a fully managed platform that makes it easy to work with. Users do not need to be concerned about hosting the software and the resources. On the other hand, Apache Kafka was initially developed at LinkedIn and is an open-source solution. A user has to take care of installation, monitoring and security of clusters, durability, failure recovery and ensuring high availability. This can be an administrative burden.

3.6.2   Cost

Apache Kafka requires hosting and managing of the framework. Total cost of the application in this depends on prudent selection of storage capabilities and computing resources which requires capacity planning. Cumulative cost of such a system is sum of resource cost and human cost. On the contrary, human resources are significantly lower given the hosted nature of Amazon Kinesis. Nevertheless, Kafka is more cost efficient in certain data ingests patterns and should be considered carefully.

3.6.3   Architectural Differences

There are few architectural differences between Kafka and Kinesis. These include scalability models, consumption latency and end-to-end data ingest. For example, in order to scale Amazon Kinesis, either a shard can be split to increase capacity or two shards can be joined to lower cost and reduce capacity, whereas Kafka requires monitoring for hot partitions and partitions can be added or moved as per the requirement.

User applications are determined by how APIs work. So, it is essential to keep in mind the features offered by either solution while choosing which streaming platform to be used. For example, Kafka enables users to replicate data based on certain key as it has the capability to keep hold of last known message. In contrast, this feature is not available in Amazon Kinesis and has to be built using API.

3.6.4   Performance

In a reasonable price, it is tough to replicate Kafka's performance in Kinesis. Kafka is faster than Kinesis. Kinesis is costly in terms of latency and throughput as it writes each message synchronously to three different data centers.

3.6.5   Summing Up

It is clear that despite being similar, Kinesis and Kafka are meant to be used in different contexts. Pros of Kafka are its performance and integration with other big data frameworks/platforms. Kafka requires in-depth knowledge to make the infrastructure less risky.

Built-in-reliability and simplicity are Kinesis's core strength. Streams only store records for 7 days and the data gets lost, some other storage is required if data is to be kept for longer. Throughput is less as compared to Kafka.

4   Conclusion

In this chapter, various concepts related to Apache Spark Streaming API were discussed. Apache Spark API provides a framework that enables scalable, fault-tolerant streaming with high throughput. In this chapter, we started with a short overview of streaming and spark streaming. After the introduction, fundamentals of Spark streaming, its architecture were focused upon. We described several transformation API methods available in spark streaming that are useful for computing data streams. These included Spark streaming API RDD—like operations and Streaming Context. We also included in our study various other operations like DataFrame, SQL, and machine learning algorithms which can be applied on streaming data. We concluded this section with discussion of steps necessitated in a prototypical spark streaming program.

Apache Kafka is a distributed streaming platform that enables processing of streams of records in the order they occur, i.e., it can be thought of a message queue which is distributed, partitioned, and replicated commit log service. In the next section of the chapter, a brief introduction of Apache Kafka and how it is used with spark, its benefits when integrated with Spark were given. We further described in detail a real-time streaming data pipeline build using Kafka integrated with Spark. We continue this discussion with another real-time parallel data processing stream known as Amazon Kinesis Streams and its integration with spark.

Finally, we conclude with trade-offs between Apache Spark and Amazon Kinesis, focusing on building a real-time streaming pipeline using available Spark API. To make the chapter more realistic and to enhance understanding of the user, wherever possible, code snippets have been provided.

References

1. Apache Spark. https://spark.apache.org/
2. Zaharia, M., Xin, R.S., Wendell, P., Das, T., Armbrust, M., Dave, A., Meng, X., Rosen, J., Venkataraman, S., Franklin, M.J., Ghodsi, A., Gonzalez, J., Shenker, S., Stoica, I.: Apache spark: a unified engine for big data processing. Commun. ACM CACM Homepage Archive New York NY USA **59**(11), 56–65 (2016)
3. Dean, J., Ghemawat, S.: MapReduce: simplified data processing on large clusters. In: Proceedings of the 6th Conference on Symposium on Opearting Systems Design & Implementation, San Francisco, CA, pp. 10–10 (2006)
4. Logothetis, D., Trezzo, C., Webb, K. C., Yocum, K.: In-situ MapReduce for log processing. In: USENIX Annual Technical Conference (2011)
5. Scala; http://www.scala-lang.org
6. Alsheikh, M.A., Niyato, D., Lin, S., Tan, H.P., Han, Z.: Mobile big data analytics using deep learning and apache spark. IEEE Netw. **30**(3), 22–29 (2016)
7. Owen, S., Ryza, Laserson S., Wills U.: Advanced Analytics with Apache Spark. O'Reilly Media (2015)
8. Spark homepage. http://www.spark-project.org
9. Apache Hive; http://hadoop.apache.org/hive
10. Balazinska, M., Balakrishnan, H., Madden, S.R., Stonebraker, M.: Fault-tolerance in the Borealis distributed stream processing system. ACM Trans. Database Syst. **33**(1), 3–16 (2008)
11. Shah, M., Hellerstein, J., Brewer, E.: Highly available, fault-tolerant, parallel dataflows. In: Proceeding of ACM SIGMOD Conference, pp. 827–838 (2004)
12. Zaharia, M., Das, T., Li, H., Shenker, S., Stoica, I.: Discretized streams: an efficient programming model for large-scale stream processing. In: 4th USENIX Workshop on Hot Topics in Cloud Computing (2012)
13. Zaharia, M., Chowdhury, M., Das, T., Dave, A., Ma, J., McCauley, M., Franklin, M., Shenker, S., Stoica, I.: Resilient distributed datasets: a fault-tolerant abstraction for in-memory cluster computing. In: NSDI Proceedings of the 9th USENIX Conference on Networked Systems Design and Implementation, pp. 2–2 (2012)
14. Apache Kafka. https://kafka.apache.org/
15. Amazon Kinesis. http://docs.aws.amazon.com/streams/latest/dev/introduction.html
16. Nair, L.R., Shetty, S.D.: Streaming twitter data analysis using spark for effective job search. J. Theor. Appl. Inf. Technol. **80**(2), 349–353 (2015)

Big Data Analysis in Cloud and Machine Learning

Neha Sharma and Madhavi Shamkuwar

Abstract In today's digital universe, the amount of digital data that exists is growing at an exponential rate. Data is considered to be the lifeblood for any business organization, as it is the data that streams into actionable insights of businesses. The data available with the organizations are so much in volume that it is popularly referred as big data. It is the hottest buzzword spanning the business and technology worlds. Economies over the world is using big data and big data analytics as a new frontier for business so as to plan smarter business moves, improve productivity, improve performance, and plan strategy more effectively. To make big data analytics effective, storage technologies, and analytical tools play a critical role. However, it is evident that big data places rigorous demands on networks, storage and servers, which has motivated organizations and enterprises to move on cloud, in order to harvest maximum benefits of the available big data. Furthermore, we are also aware that conventional analytics tools are incapable to capture the full value of big data. Hence, machine learning seems to be an ideal solution for exploiting the opportunities hidden in big data. In this chapter, we shall discuss big data and big data analytics with a special focus in cloud computing and machine learning.

Keywords Big data · Big data analytics · Machine learning · Cloud computing
Cloud technology · SaaS · PaaS · IaaS · Supervised learning · Unsupervised learning

N. Sharma (✉)
Vidya Pratishthan's Institute of Information Technology (VIIT, Baramati), Pune,
Savitribai Phule Pune University, Ganesh Khind, Pune 411007, Maharashtra, India
e-mail: nvsharma@rediffmail.com

M. Shamkuwar
Zeal Institute of Business Administration, Computer Application and Research, Savitribai Phule
Pune University, Ganesh Khind, Pune 411007, Maharashtra, India
e-mail: madhavi.sh@gmail.com

© Springer Nature Singapore Pte Ltd. 2019
M. Mittal et al. (eds.), *Big Data Processing Using Spark in Cloud*,
Studies in Big Data 43, https://doi.org/10.1007/978-981-13-0550-4_3

1 Introduction

When the data grows large enough in all perspectives like volume, velocity, variety, value, and complexity so that it becomes difficult to process it with traditional approaches, then it is certainly a big data [1]. Big data is generated when the volume of the data increases from terabytes to zettabytes, velocity increases from batch data to streaming data, when variety, value, and complexity increases from structured data to unstructured data and when existing database management tools become incapable to record, modify, query, and analyze [2, 3]. Big data analytics use modern analytic techniques which help businesses to take quick and agile decisions based on a 360° view of the organization and its various operational processes. Furthermore, if companies improve the quality of data, its usability, intelligence applied, remote accessibility and sales mobility, it can easily generate profits as well as have the cutting edge over the competitors. However, big data poses both opportunities and challenges for businesses. The opportunities lies in unveiling the benefits in different sectors of business and choosing it as a career of a technology professionals who will explore and exploit the power of big data. And the challenges are due to the exponential growth of data that places a high demand on storage systems and infrastructures, which are incompatible in terms of software and hardware needed to assimilate and process big data.

Cloud technology is very popular in medium- and small-scale business as it provides cost-effective and lower risk solutions and is considered as a broader term for most of the online services. Companies which do not have the in-house skills to perform analytics on aggregated unstructured data are turning towards cloud for Software as a Service (SaaS), Platform as a Service (PaaS) and Infrastructure as a Service (IaaS) solutions as they are the fastest growing delivery models. The elasticity of the cloud makes it ideal for analysis of structured and unstructured big data. Real-time analysis is allowed by the cloud technology to explore meaningful information from the big data which is stored on the cloud platform in order to reduce the cost of storage. Presently, the trend in the industry worldwide is to perform big data analytics in cloud, as the duo provides high scalability, agility, affordability, and many more benefits.

The process of performing analysis of a very large size complex data to explore hidden patterns which exists in it but not yet revealed, with the objective to facilitate decision-making, is termed as big data analytics. So far, the existing traditional tools are not efficient enough to seize the whole potential of big data, therefore there is a need to integrate machine learning to big data. Machine learning is the science of creating algorithms and programs that can learn on their own. It thrives on growing data sets which is disparate in nature and has large set of variables, unlike traditional database which is small, simple, and of homogenous nature. The result of machine learning system improves with the quantity of data fed, as with these systems there is no limitation of human scale thinking and analysis, hence it helps the system to learn more and present deeper insights.

In this chapter, we have attempted to give a brief insight of big data and its inter-connection with cloud technology/computing and machine learning algorithms/mechanism. This chapter is arranged as follows: the next section reviews the work done by researchers with similar interest; Sect. 3 gives the complete information about big data from all the perspectives; merger of big data and cloud technology is discussed in Sects. 4 and 5 covers the significance of machine learning for big data. The chapter would have been incomplete without presenting challenges and opportunities related to big data and analytics, hence discussed it in Sect. 6. Finally, Sect. 7 summarizes the chapter with conclusion, followed by references which are mentioned in the last section.

2 Related Work

The following section provides literature review of big data analytics which helps the stakeholders to make appropriate strategies and develop business processes in accordance with the outcome of analysis. The challenges and opportunities of big data regarding various market segments are discussed in [4]. An in-depth study on emerging technologies such as Internet of things, social networks, smart grid are given in detail in [5]. Furthermore, data management challenges are data collection, data aggregation, data diversity, data variety, data reduction, data integration, cleansing indexing and tools for data analytics are shared in [6]. A big data processing model with the view point of data mining is proposed in [7]. The issues for big data such as storage infrastructure, transport mechanisms were discussed and a detailed study on what could be the future research area of big data computing was conducted [8, 9].

A detailed study on big data architectures and framework are specified by the National Institute of Standards and Technology Big Data Public Working Group [10]. The several areas influenced by implementing big data computing are described in following section:

Scientific Explorations

The experiments pertaining to physics, biology, astronomy, medical sciences, and social sciences are collected, analyzed, interpreted, and visualized for the benefit of mankind. It provides services in various sectors such as consumer goods, banking, telecommunications, life sciences, health insurance, sports, travel and transportation, retails, health care, Internet, manufacturing, education, weather prediction, prevention from natural disasters, etc.

Health Care

Big data analytics have increasingly gained its importance in the healthcare industry as it helps in estimation of number of incidences of diseases. It is used to detect diseases at its early stage, treats the patients effectively, and prohibits the disease from further spread [11]. A collection of files from websites, survey reports, mails, patient feedback forms, discharge summaries, etc., can be analyzed to determine the number of incidences of the disease. Certain outcomes may be predicted or estimated based on historical data such as length of patient stay, which patients that would choose what type of surgery (elective surgery) or hospital acquired illness and its progression.

Surveillance System

The system includes energy and utilities, communication and government sector. According to Public Sector Research publication (PwC), the use of more big data analytics is more prominent. Increase in connectivity of infrastructure in government sector fuels the deployment of big data analytics tools and prevention solutions. For example, SAS Institute Inc. has developed government analytics solution to detect fraud and prevent improper payments across variety of areas such as tax and revenue, procurement fraud and unemployment and workforce programs. Additionally, governments are adopting various authentication systems and solutions for users such as single sign-on and multifactor authentication for reducing unauthorized access to services. Increasing online data and electronic transactions among organizations are expected to create numerous growth opportunities for the fraud detection and prevention market in future. The data generated in the form of audios and videos are used for security and are used for security and surveillance system.

Retail Industry

Players of the retail industry are an integrated part of the supply chain of a product or service. The growth of this sector is due to a significant amount of customer feedback. To categorize and summarize the structured and unstructured data, retailers use an analytical technique. Usually big data analytics is popular in retail industry as the feedback from the customers is in manual format and text analytics help in analyzing it automatically in an accurate manner. Retail organizations would benefit from the data provided by big data analytics as it would help in gaining valuable insights, which would improve the business process, help in determining customer interests and needs and helps in reducing the number of product returns resulting from poor quality. The retailers believe that there is lot of valuable information hidden in the unstructured data and they need techniques which will grant them access to that information. Text analytics is being used by the retailers to gain insights from consumer satisfaction surveys, contact center transcripts, and Internet sources. Active Point Crossminder and Alceste are some tools used for analytics that offer natural language processing and analysis of all kinds of texts. Confirmit Genius Analytics enables retailers to gain deep insights about their customers. It delivers a high-quality analysis that would help the companies to understand their customer's perspective regarding the business. The analysis of the real-time data into sentiments and to identify key issues and trends is enabled through Confirmits' text analytics solutions.

3 Big Data: An Inside Look

Increase in the number of mobile devices and applications as well as organization's shift from analog to digital technologies has driven the market for big data and big data analytics. Big data is the term not only given to tremendous amount of data and large data set but, the overall technology which deals which such massive amount of data, whereas big data analytics is the process of analyzing large data to expose hidden patterns, understand customer preferences, explore unknown correlations and other useful information pertaining to business. In highly regulated sectors like finance, healthcare, insurance, retail and social security, combating fraud is an essential, as there is large amount of compliance, regulations, risk management measures, and monetary consequences to deal with. Hence, large enterprises in private and government sectors are deploying big data solutions to manage the data generated, to provide new insights and to assist in quick decision-making process.

3.1 Definition

Roger Magoulas from O'Reilly media coined the term "Big Data" in 2005 [2]. Simultaneously, there has been lots of brain storming in academia as well as industry regarding the definition of big data. However, to reach to a consensus is slightly difficult [11, 12]. The term big data is defined in various ways by considering several dimensions regarding its overall process. There are three possible definitions of big data are described as follows:

Big Data Definition from Attribute's Perspective

Big data is described by five V's, i.e., Variety, Volume, Velocity, Veracity, and Value. The volume of big data indicates the data as rest, which is the opulent quantity of data collected over the last few decades and is continuously being captured every moment. The profuse amount of data is gained with a simple reason that no data is considered as insignificant and hence saved for future references. The variety of data indicates its various forms like structured, unstructured, text and multimedia. Any data form for which there is well-formed set of rules formed is called as structured data such as employee's database, ultra-sonography reports, blogs, audio, video, etc. When there is the absence of such set of rules, it is termed as unstructured data such as comments, tweets, and feedback given for online shopping. In such cases, the sense of the data needs to be extracted by the use of technology. The velocity of data refers to the data in motion like streaming data. Frequency with which the data is generated at every tick of the clock, right from SMS from friends, social media posts, card swipes, email updates, financial transactions, promotions messages, etc., to the machine/sensor generated data, everything adds to the velocity of data. The value of big data denotes the costs and benefits of aggregating and analyzing the data to ensure that ultimately the data that is reaped can be monetized. Veracity refers to the data in doubt and indicates towards the need of reliability and trustworthiness in the data. The data is meant to be free from all discrepancies such as incompleteness, outliers, data redundancy and inconsistency. These five V's are considered as the pillars of big data [13–17].

Big Data Definition on the Basis of Comparison

The study by McKinsey suggests that any dataset which surpasses the capability of usual database management system tools and techniques to capture, store, manage, and analyze [3]. As compared to attribute definition, where we find a direct definition in terms of five metrics, the current definition is highly subjective. It does not suggest the size of the data to qualify for big data.

Big Data Definition on the Basis of Architecture

The National Institute of Standards and Technology (NIST) defines big data as a dataset whose volume increases with the velocity that it cannot be processed by traditional relational concept and hence significant horizontal scaling is needed for efficient processing [18]. Therefore, it is suggested that big data should be further categorized into two specializations as mentioned below

1. Big data science	It encompass study of techniques involving the collection, cleansing, and estimation of big data
2. Big data frameworks	It includes software libraries and related algorithms that facilitates processing in a distributed manner as well as analysis of the problems in big data

3.2 Paradigms Shift: Traditional Data to Big Data

Data management and analytics methods have seen a historical change in the last decade due to the drastic increase in the use of internet, its wide-spread applications and significance of data known recently, hence resulting in a paradigm shift from traditional data to big data. Traditional databases accept limited amount of structured data usually stored in relational databases or data warehouses. The relational databases performs online transaction processing that causes frequent changes (insert, update, delete) to the data, whereas data warehouses are mainly meant for analytics through online analytical processing or data mining to support process of decision-making. Today the Big data has evolved as the backbone of businesses and world economy as it denotes scalable data architectures and effective tools to manage large volume of unstructured data generated in a high speed. Meanwhile, the data complexity and its size on the network are on a rise; thus, increasing the concerns about the vulnerability of big data application. The big data concept has become popular not with blink of eyes, but it took several years to jot down the need

Table 1 Comparison of traditional data and big data

Differentiating parameters	Traditional data	Big data
Volume	Gigabytes to terabytes	Hundreds of petabytes of data or zettabytes of data and beyond
Velocity and throughput	Slow and lower	Rapid and higher
Variety (Data type)	Structured data and semi-structured data	3D data, geospatial data, video and audio, structured data, semi-structured data, unstructured data like log files, sensor data, social media data
Data source	Centralized	Distributed
Data integration	Simple	Complex
Data store	RDBMS, SQL	HDFS, no SQL
Access	Interactive	Batch or real time
Architecture	Centralized	Distributed
Schema	Fixed	Dynamic
Scalability	Low	High
Costs	Use of expensive hardware and software systems	Less expensive as it makes use of commodity hardware and open-source software
Accuracy and confidentiality	Less	More
Platforms	Relational database management systems, online transaction processing	Big data analytics, web log mining, intrinsic information extraction, text mining, social networking, video analytics, graph analytics, in-memory analytics, scientific data exploration, statistical analytics and predictive analytics
Theorems	CAP theorem [19] ACID properties [20]	CAP theorem [19] BASE properties [21]

of this concept, development of concept and applicability of tools. Comparison of traditional data and big data is presented in very comprehensive way in Table 1.

3.3 Evolution of Big Data

The stereotypical nature of data and related functions was limited to its collection, processing, and display. However, the technology always provided supporting spine for the applications to depend on. The earlier descriptions of the data were based on traditional timeline based order or in form of milestone technological achievements

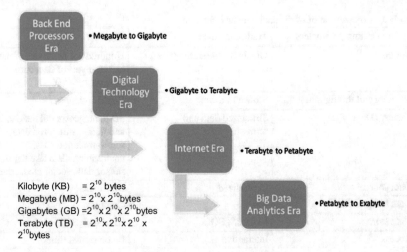

Fig. 1 Evolution of big data according to data size from megabyte to exabyte

[22, 23]. In this sub-section, the history of big data shows the variations of data sizes as per the requirements, technology evolution, infrastructural development, and advancement in operating environment. The change in technology with respect to growing time is shown in Fig. 1. Various era of big data evolution is discussed in following texts.

Back End Processors Era

In the year 1970s and 1980s, conventional business data introduced the concept of storing, handling, and managing massive amount of data by showing a giant leap in data size from megabyte to gigabyte sizes. The technological need was to have a data container which will hold the data and handle queries so as to retrieve data in most efficient way. As a result, back end processors were made as a combination of specialized hardware and software, with an intention that it would perform better and be cost effective.

Digital Technology Era

Until late 1980s, data storage and retrieval was not a big issue, due to extensive use of back end processors, however, further to improve the system performance, data parallelization concept was evolved. The set target for performance enhancement was ensured by implementing shared disks [24], shared memory, shared nothing, shared processors and hierarchical databases as per the need or architecture. The databases distribute/retrieve the data to/from multiple storage and performs database indexing. The commercialized products which have adopted such technology are Teradata, Netezza, Greenplum, Aster Data, and Vertica [25–29]. The data exceeded from gigabyte to terabyte in the digital technology era.

Internet Era-The Fourth Paradigm

The big bang advancement in late 1990s is Web, which is termed as "the fourth paradigm" [30] by a database and software pioneer—Jim Gray. The technology not only changed the transaction style, working style but the way businesses were carried throughout the world. The need to store structured, semi-structured, and unstructured data led to efficiently hold the data from terabytes or petabytes. As parallel databases were fully developed to its mature state and were unable to handle such complex data and data size beyond terabytes, therefore Google created Google File System which is also known as GFS or GoogleFS and MapReduce [31, 32]. GFS manages the web spawns load of data generated from user content and MapReduce is a programming model which provides distributed data storage and automatic data parallelization over the clusters of hardware. Both the technology provides high degree of scalability and process unlimited amount of data.

Big Data Era

In recent years, social media and e-commerce are dominating the industry and ruling the market worldwide. Introduction of big data analytics, cloud computing services, and increase in cashless payment are strengthening the data growth. The companies adopt efficient data analytics techniques and cloud technology to analyze the plethora of data available through different source, in an effort to provide customers with innovative and better service. The ability of big data analytics to evaluate the real-time data has motivated the businesses to adopt it; hence, it is achieving its reputation as a revolutionary technology that has a power to redefine business processes.

3.4 Big Data Competency Taxonomy

The capability of big data analytics to derive meaning or sense from unstructured data, certainly benefits the companies in attainment of business intelligence, realizing market subtleties, and gaining significant insight. The big data competencies which can be classified into six elements on the basis of key aspects needed to establish big data analytics infrastructure are technologies, tools, data dimensions, programming models, security, analytics, etc., as shown in Fig. 2. The big data taxonomy provides an in-depth study of factors used to analyze customer behavior, purchase patterns, analyze the needs, demands or interests, etc. The brief about each dimension is discussed in the following texts.

3.4.1 Technologies

The major technologies used for big data are file system, open-source frameworks, and commercial frameworks. The brief information about the big data technologies are discussed below

Big Data File Systems:

A file system performs number of operations like storing and retrieving, organizing, and naming the files and folders along with safety and security of file systems with

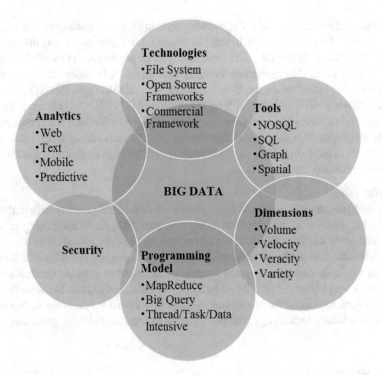

Fig. 2 Big data competency taxonomy

the help of local as well as distributed file system. The major challenges of big data file system are read/write operations, concurrent access, synchronization techniques among multiple applications, request for creating file system, and data transparency. The designing goals of the file system should include

- The distributed form of big data file management system should provide distributed access and at the same time the files are treated as local files.
- The file system should provide remote file access without providing any local information.
- The file system should efficiently handle failures occurred in application programs and client side programs.
- The heterogonous nature of file systems should handle diverse software and hardware components.
- The efficient file system of big data should be tolerant towards network partitioning which allows data to be available, in case of failures.

Big Data Open-Source Computing Framework:

The information regarding big data open-source computing framework is given below

- Hadoop functions on a distributed processing technology and is used for big data analysis. In recent times, the Hadoop market has grown at exponential rate due to its efficiency and as it is financially feasible, as compared to RDBMS and other contemporary data analysis tools [33].

- Storm and S4 are designed to work on stream data. Storm is a designed for performing real-time complex analytics on stream data using distributed Remote Procedure Call (RPC), machine learning, continuous processing and ETL (Extract, Transform and Load), whereas S4 is designed to management of the stream data [27, 28].
- Spark is highly scalable, fault tolerant framework which provides resilient distributed datasets (RDDs). Spark is a general engine which provides high performance especially for large-scale data processing. The tools included are Shark SQL, graphical analytics tools, Spark Streaming, MLib machine learning, and Graphx [34].

Big Data Commercial Frameworks:

The information regarding big data open-source computing framework is given below

- CloudComet is an autonomic computing engine provides infrastructure and programming at run time to provide scalability for performing online risk analysis. CloudComet may use both public and private Cloud provider to handle extensive demands of online risk analytics.
- In February, 2014, Oracle acquired BlueKai, a cloud-based big data platform management company to improve its data management in the cloud. In addition, Bluekai had also personalized Oracle's online, offline and mobile marketing campaigns.
- Google has a commercial framework to perform big queries on Google big tables [35, 36].
- Amazon flexible MapReduce is a managed Hadoop framework to process big data across dynamically scalable Amazon EC2 instances in very efficient manner [37].
- HDInsight is a Hadoop framework of Windows that runs on the Microsoft Azure platform [38].

- RackSpace, Aneka, Horton, and Cloudera also offer commercial frameworks to compute and process big data [38–42].

3.4.2 Tools

Big data tools that help to sort through data are discussed in the following texts:

NOSQL Technology

NoSQL is an abbreviation for "Not Only SQL" is the framework of databases especially designed for handling massive amount of data and performs agile processing. It is adaptable to the ever-growing demands of big data. This database infrastructure handles structured and unstructured database stored at physically different locations using key-value pair table (KVP). The table has two columns to store keys and values and are best applied on Hadoop MapReduce environment and adopted by Amazon's Dynamo and Oracle's Berkeley DB [43, 44].

SQL-like Query

Structured Query Language (SQL) is a query language used to create, store, and manage traditional databases. Such databases are also a part of big databases like document-oriented databases which contains documents encapsulated in standard formats. 10gen's MongoDB and Apache's CouchDB are the examples of such databases [45, 46].

Graph Big Database

Graph database processes complicated large-scale many-to-many and multi-level relationship by adopting graph properties and graph structures (nodes and edges). Graph database does not need indexes but makes use of pointers to point to the adjacent elements. Neo4j open-source graph database and objectivity's infinite graph are the examples of graph databases.

Spatial Databases

Spatial databases are an important big data tool. OpenGIS (Geographic Information System) established by Open Geospatial Consortium (OGC) has also taken efforts to standardize spatial data. The spatial databases contain the elements like lines, points, and polygons. Spatial indices work on the pictorial representations found in any database and supports ad hoc queries.

3.4.3 Dimensions

As discussed in above section, there are five dimensions of big data, which are popularly known as five V's, i.e., Volume, Velocity, Veracity, Variety, and Value. Each dimension is discussed in the following texts:

Volume

Sources like Internet, mobile phones, digital devices, RFID sensors, social networking sites, satellite images, etc., are liable for the data deluge, which is estimated to surge at a rate of 40% every year at global scale [47]. It is clearly evident from IDC report that the volume of global data has doubled every 2 years from 2005 and will further grow at the rate of 400 from now to reach 40 zettabytes by 2020 [48].

Velocity

The speed with which the data is generated can be attributed to mandatory online transactions in private as well as government sector, growing business intelligence applications, and rise in the use of sensors. The increased velocity of data collection is a result of improved Internet connections and its usage.

Veracity

The data used for big data analytics should be free from anomalies so as to be used for decision-making. Therefore the big data has to undergo filtration and cleansing process to select the relevant data for analytics.

Variety

Big data comprises of all the types of data like structured data (data stored in relational database), semi-structured data (data stored in XML documents), and unstructured data (data in the form of video, audio, text, etc.).

Value

Value of big data denotes the quality of data stored, as it is expected that the big data will benefit the businesses through the analytics and ensure financial growth along with overall prosperity.

3.4.4 Programming Models

Different programming models used to compute and process big data are discussed below

Map Reduce

It is a high-level data-intensive programming language which performs Map and Reduce functions. Map function filters and sorts the data, whereas Reduce function performs aggregation of Map output, and finally generates the final result. The examples of Map Reduce programming models are Apache Spark, Hadoop MapReduce, and Aneka MapReduce [27, 49, 50].

Big Query

It is the query language for big data and is mainly used while performing weblog mining to study the data on web. Google big query is an example of big query.

Thread/Task Data-intensive

The thread data-intensive model is preferred for computing logic and usually used for high-performance applications, whereas task data-intensive model is preferred for workflow programming.

3.4.5 Security of the Big Data

Big data ecosystem faces privacy and security challenges towards infrastructure security, data integrity, data management as well as privacy and reactive security. Big data infrastructure can be secured by distributed computations and data stores. Data can be preserved through cryptography and access control at each level. Security while managing big data essentially requires distributed and scalable solutions which is also critical for investigations of data attribution and efficient audits. Lastly, checking the data integrity of streaming data from distinct points is essential, and can be used to execute real-time analytics in order to prevent infrastructure attack.

3.4.6 Analytics

In the last few years, the importance of analytics has increased drastically. Organizations use analytics to understand their customer in a better way, analyze the customer interests, identify brand loyalty, verify reviews and feedbacks of the products in the market, etc. The process of integrating traditional analytics techniques and mathematical models is called as analytics. Analytics uses statistical models, data mining techniques, and technologies in order to analyze the massive data. Analytics and analysis are the term used interchangeable, however, analytics is scientific application of analysis [51]. Big data can be analyzed in various ways such as by clustering the data into similar types, performing text analytics, analysis on real-time data, etc., hence few areas of analytics application are discussed in this sub-section.

Web Analytics

Web mining is another segment of data mining which is growing at a faster pace, where real-time data which is available on internet is analyzed. The increasing usage of social networking sites and rising awareness about the benefits of analyzing these websites on real time has raised the popularity of cloud technology among the analytics companies. Social websites are one of the common platforms of communication about various offerings (products/services) by the companies amongst the customers. Analysis of these social sites gives immense information to the companies about their customer behaviors, purchasing patterns, customer interests, brand awareness, loyalty amongst the customers, etc. This analysis enhances the company's decision-making ability, develop new marketing strategies and improve their businesses in future. Often companies feel that they do not have in-house skills to aggregate and analyze unstructured data. Therefore, many vendors are turning to cloud solutions that make the cloud technology prominent in the analytics market.

Text Analytics

Text analysis is the process of deriving high-quality information by analyzing text. The classified data is analyzed using different software tools. Text analytics software assists the companies to understand their customers better from all perspective. The software tools are provided by different players in the market such as Microsoft, IBM, Oracle, Tableau, SAS, Attensity, etc. Predictive analytics and web mining are the major elements of text analytics.

Mobile Analytics

Currently, the mobile technology is the hottest technology in business intelligence field. Various vendors are providing text analytics on mobile devices. For example, Lexalytics engine ported to Android provides native analysis within the Android devices. Also, an auto classification Web app by OpenText is designed for the processing in the mobile devices. SAS's analytics output for mobile business intelligence is supported by mobile devices and Attensity's tablet application analyzes text using a smartphone. Therefore, rise in the adoption of text analytics on mobile devices would drive the text analytics market in future. It is been noted that, at the end of 2012, mobile data traffic was 885 PBs per month [52]. This massive size of data is due to extensive use of mobile apps and multimedia available and drives mobile analytics.

Predictive Analytics

Predictive analytics help in finding patterns in data that in turn helps decision makers or in the improvisation of business processes. The information that is extracted from the analysis of data enables to plan appropriate strategies for the enterprises. The use of statistical tools and modeling techniques to estimate the future trends depending on the current and historic data is predictive analytics. Predictive analytics create a mathematical model to use as the basis of predictions and provide an insight into the characteristics and behavior of the buyers. It checks for the pattern or the trend in the past that are likely to emerge again, which allows businesses and investors to have a futuristic outlook. There are various predictive analytic models available in the market that analyze the future trends by analyzing past trends that are likely to emerge in future. However, descriptive models use past data to determine the response of a group to a certain set of variables. Primarily, predictive analytics are the decision-making tools for various industries. For example, marketing department analyzes the consumer behavior according to the past trend in the market and predict the future demand for a particular product. Predictive analytics helps the organizations to forecast the probable future with adequate levels of reliability and plan risk management strategies.

3.5 Big Data Journey: Lifecycle Phases

As any system has a life cycle, so do there is a life cycle of big data, which deals with the four phases of digital data, from its birth to its destruction. The following section throws a light on four phases of big data life cycle as per systems-engineering approach [53, 54]. The digital data life cycle includes big data generation, big data acquisition, big data storage and big data analytics as shown in Fig. 3.

Phase 1: Data Generation

Due to advancement in technology and lowered price of storage devices, the data generation rate is a notable trend these days. Cisco holds fast popularity of audios, videos, use of Internet and social networking sites responsible for generating the new data [55]. As per estimation of IBM, 90% of data is generated in last 2 years [56]. The main reason for this data deluge is ICT, which makes data available for storing, processing, reading, etc. The data generation is accomplished in three different stages, which are sequential in nature. These stages are explained in Table 2.

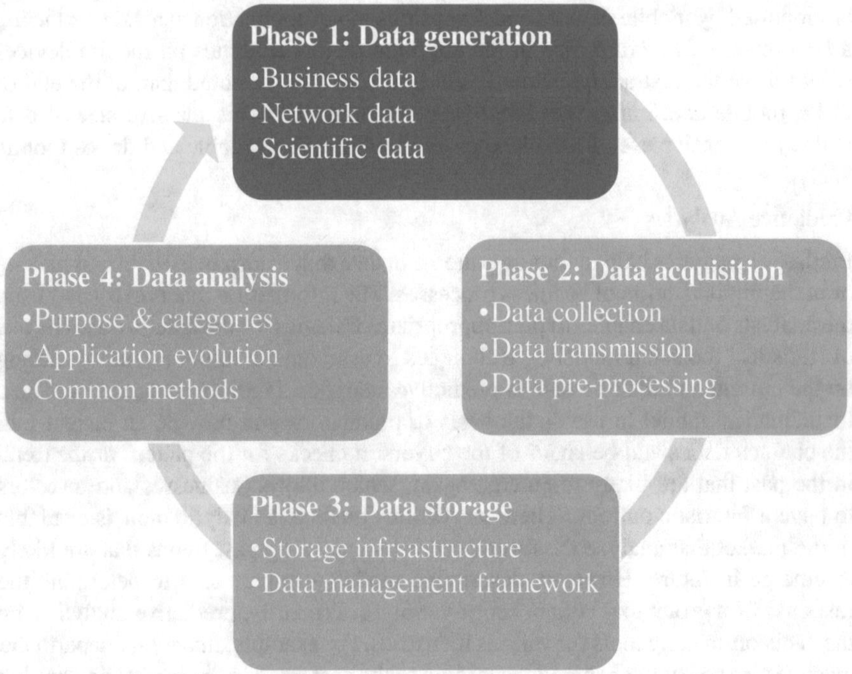

Fig. 3 Life cycle of digital data

It is quiet illustrative that in the near future, the automatic data generation will contribute towards data collection in a big way. Thus, rate at which data generated will be high and time required to generate data will be low. Moreover, the large volume of data is generated by different domains and is capable of creating potential value. Following are the three domains which are explored further, and vary in terms of its conception and over all execution.

Business Data

Businesses are contributing more to the growth of digital data and are continuously amplifying its power. As per the projected value, the volume of data is doubled every 1.2 years in the businesses around the world [57]. It is also estimated that B2B and B2C business transactions over the internet will touch 450 billion per day [58]. The speed of data generation is more effective and demands real-time system to perform analysis. As per sources, millions of queries and back end operations are handled by Amazon on daily basis [59]. Similar is the case with Walmart which handles more than a million transactions per hour which contributes to over 2.5 PB's generation of data [60]. Akamai, is the leading content network which analyses handles 75 million events per day, so as to get better target advertisements.

Table 2 Stages of data generation

Stage	Entry time-line	Technological implications	Sectors influenced	Data type	Data genera-tion
First	1990s	• Extensive use of digital technology, • Analysis done through database management systems	Archives from government, shopping records, bank trading etc.	Structured	Passive record-ing
Second	Late 1990s	• Extensive use of web search engines and e-commerce business • Data generation by user generated data	E-commerce and social networking	Semi-structured and unstructured data such as blogs, discussion forums, social networking	Active genera-tion
Third	2010	• Extensive use of electronic devices and mobile devices • Dependent on mobile centric network which is location centric, person centric and context relevant	Electronics and telecom sector	Semi-structured and unstructured data	Automatic genera-tion

Network Data

In the daily livelihood of human beings, Internet, Internet of things, wireless sensor networks are making impact. The data is bombarded and requires advanced tech-nologies to handle it. In 2008, Google, the web search engine used to process 20 PBs of data [32]. Facebook—a social networking platform stored, accessed, and the analyzed over 30 PBs of content generated by user. Twitter—online news and social networking platform generates more than 32 million searches per month. In 2010, 4 million people used mobile phones and 12% peoples have smart phones, adding a heavy load to mobile networks. The other sectors such as transportation, retails, industrial, automotive the number of sensors increases 30% per year [17].

Scientific Data

It is also noted that the scientific data are being generated in large quantity which belongs to several disciplines. It led to a need to analyze this large volume of data. The scientific domains which use big data analytics are discussed further.

Computational Biology: GenBank is a database responsible to store nucleotide sequences and is maintained by National center of Biotechnology. It is estimated the database doubles in every 10 months. In the year 2009, GenBank consists of 250 billion records belonging to 150,000 distinct organisms [61].

Astronomy: Sloane Digital Sky Survey (SDSS) is the largest astronomical catalog which had generated over 25 TBs of data from telescopes, in the span of 10 years, i.e., from 1998 to 2008. As the resolution of the telescope is increased for good image quality, it is estimated that over 20 TBs data will be generated in the year 2014 [62].

High energy Physics: Centre for European Nuclear Research worked on Atlas experiment which is a general-purpose detectors at the Large Hadron Collider (LHC). This experiment generates 2 PBs data in 2008 and the data generation rate is about 10 PBs per year [63].

To participate in analyzing the data, geo-distributed parties needs to collaborate with each other, as the huge amount of data is generated [64, 65].

Data Attributes

The data generated is collected from heterogeneous data sources and are complex in nature. The diverse data types are spatial, temporal, scale, dimensional, etc. NIST had defined five attributes which characterizes the big data as mentioned below

- Volume defines the sheer volume of datasets.
- Variety defines different forms of data such as structured, unstructured, and semi-structured data form.
- Velocity describes the speed at which data is generated.
- Relational limitation includes two types, i.e., temporal and spatial data.
- Horizontal scalability defines the coordination of individual resources and ability to join multiple datasets.

Phase 2: Data Acquisition

The big data analytics accumulates the data for storage and analysis. The data acquisition step in turn consists of three sub-steps: aggregation of data, transmission of data, and preprocessing of data. The latter two phases do not follow any strict order, data transmission can be done after or before the data preprocessing steps.

Data Collection

A well-designed process which helps to fetch the data from multiple sources is called as Data collection. The method of data collection process takes into consideration not only physical characteristics of data sources but also the objectives to build data sources. The data collection is quiet sensitive as inappropriate data collection methods will result in accumulation of invalid results. The data collections methods are of two types, the pull based approach where the data is actively collected by a centralized or a distributed agent. The other method is push-based approach, where the data is pushed from source towards the sink with the help of a third party. The three methods for data collection are given below

(a) *Sensors*

Sensors are the wired or wireless networks meant to note the physical characteristics such as temperature, pressure, humidity, moisture, chemical, acoustic, sound, vibration, etc. these physical characteristics are converted into digital signals and then stored as well as processed. WSN is gaining popularity due to its multiple applications in diverse fields such as military, environment research, space exploration, vehicular movement, factory environment monitoring, disaster relief applications, water monitoring, forest fire detection, remote patient monitoring, civil engineering, temperature monitoring, wildlife monitoring, [66, 67] etc.

(b) *Log File*

Log file is one of the popular data collection method used in digital devices, which records activities and events in specified file formats for its analysis. A web server records the user activities such as clicks, hits, access, downloads [68]. The log file formats available are in ASCII format, the log file formats are given below

- Common Log File Format by NCSA
- Extended log format by W3C
- IIS log Format by Microsoft

To improve the efficiency of gigantic log repositories, databases are used instead of text files to store log [69, 70]. In stock markets, network monitoring, traffic management log file based data collection is used. The log file method is coined as "software as a sensor", which uses certain algorithm to sense the activities performed by the user and system [71].

(c) *Web Crawler*

The web crawler is a software program which facilitates the downloading and storing of web pages for a web search engine. First, the web crawler starts with a preliminary set of URLs to be visited which are marked in a queue. Depending on the web crawler policies, certain web pages with a given priority is visited, there after downloads corresponding all the related URLs in the web page are then add the new URL in the queue. This iteration continues until the crawler decided to halt the crawling process.

The entire crawling process is done on the basis of certain policies such as [72]:

Selection policy: It communicates the crawler about pages to be downloaded

Revisit policy: The policy decides when to check for changes in pages

Politeness policy: The policy from overloading the website

Parallelization policy: The policy is responsible to establish coordination among distributed web crawlers.

The internet applications are becoming richer and richer day by day, in order to be compatible with them, many crawling strategies are being formed [73].

Data Transmission

After data collection is done, the data needs to be transmitted into a storage infrastructure. After storage, the data is processed for its further use. There are two prevalent processes for data transmission.

(a) *IP backbone*

To transfer big data from its source to data center, a high capacity trunk line is provided. The physical media and link management techniques are used to determine factors like data transmission rate and capacity.

Physical Media: To increase the data carrying capacity, many optical fiber cables which are bundled together are used. In case of failure, the physical media must guarantee the path diversity to reroute the data traffic among multiple channels.

Link Management: It deals with the way in which the signal is transmitted over the physical media. IP over Wavelength Division Multiplexing (WDM) is developed and being used since last decade, which multiplexes on a single optical fiber to carry data signals [74, 75]. Orthogonal Frequency Division Multiplexing (OFDM) provides data transmission flexibility, high speed data transmission, agility and is resource efficient in nature.

(b) *Data Center Transmission*

Data center transmission deals with the process of placement adjustment and processing of data collected at data center. The data center network architecture and transportation protocol are closely related with data center transmission and are explained in the succeeding texts.

Data Center Network Architecture: These data centers are collection of racks over which pool of servers are kept. They are connected via data center's internal connection. The topology for the data center networks should be very efficient. The demand of maximum bandwidth and reduction in energy consumption is a challenge while discussing about shortage of electronic packet switches. Nowadays, data center network with optical connection are widely used as they provides with higher throughput rate, lower latency, and lower energy consumption. Optical technology provides high energy efficiency and provides data transmission in tbps. Recently, many optical interconnect schemes are being implemented which either upgrades the network or replaces current switches [76–81].

Transportation protocol: When large amount of data needs to be transferred TCP and UDP protocols are not efficient enough. Therefore, enhanced TCP protocol is used with an objective to enhance the throughput of the link at reduced latency.

Data Preprocessing

The data sources are heterogeneous in nature, hence the quality of data produce vary in terms of formats, noise, duplication, consistency, etc. So, to transfer the data and to store it further will costs in terms of processes and time required. Data analysis strictly requires data to be of good quality. Data preprocessing and data federation techniques are applied to improve the data quality and thereby contributing towards quality analysis. There are three data preprocessing techniques which are used to improve the data quality are discussed below [82, 83]:

(a) *Integration*

The integration is performed to provide the user with a unified view of data which is generated from collection of heterogeneous data sources [84]. The traditional approach followed by data warehouse consists of an operation called an ETL-Extract, Transform and Load, and the same is adopted in the case of big data.

Extract: This step identifies the data source, selects the relevant data, and further collects the data for analysis.

Transform: This step applies set of pre-defined rules to the data and further converts it into the desired standard format.

Load: This step loads the data generated in standard format to the storage devices.

The data federation method aggregates the data from diverse sources and further creates virtual database. This virtual database consists of data about data, i.e., metadata. This both methods work on the principle "store and pull" which is not efficient enough to provide high-performance analysis.

(b) *Cleansing*

After data collection, the data flaws such as incompleteness, inaccuracy, noisy, and outlier data are removed to improve data quality. Data cleansing techniques are must to keep data clean, free from all flaws and updated [85]. Format checks, limit checks and completeness checks needs to be completed in this process. The basic steps required to complete general framework of data cleansing are given below

- Error defining and determination
- Locate the error instances
- Error correction
- Documenting error types
- Mapping the data entry procedure with error instances to reduce future errors.

Radio-frequency identification (RFID) technology is used in inventory control system for object tracking. The technology possesses certain limitations such as low data quality, data limitations and is sensitive for environment noise. To address the missing data problem in the mobile environment, a probabilistic model is developed [86]. BIO-AJAX framework takes into consideration the biological data and standardizes so that the quality can be enhanced, and errors and duplication in the data set will be easily eliminated. Thus, data cleansing is the quality enhancement process and it largely depends on the complex relationship structure among data set. The data cleansing process should be compatible with the type of data flaws.

(c) *Data Redundancy*

Data redundancy occurs when the same set of data is placed at multiple different locations. The overhead of transmitting the data can be minimized if the data is free of redundancy. Furthermore, there is a cost associated with storing the redundant data, which can lead to inconsistency while data updation. To minimize the data redundancy, the reduction techniques are redundancy detection and data compression which gives additional task of compressing-decompressing the data apart from giving several advantages [87, 88].

In these recent days, use of camera is more prominent, which contributes to the growth in the amount of images, audio, and video. The redundancy in the form of temporal data, spatial data, statistical data, and perceptual data is more. As video consumes more storage space, video compression techniques are required for handling diverse video standards such as AVI, MP4, FLV, RealMedia, etc., are emerging to reduce the video size. Another data redundancy technique is data deduplication which eliminated the duplicate copies of redundant data and compresses the data. The next technique is storage deduplication technique which maintains an identification list which and identifies the data allotted in a segment called as hashing. The deduplication technique is useful for massive amount of data such as big data.

As discussed there is no single process or technique to perform data preprocessing, it depends on the type of data and objective behind data analysis and other factors contributing for data analysis.

4 Merger of Big Data with Cloud Technology

When big data computing is done over clouds it is known as "Big Data Clouds". To build an integrated platform suitable for quick analytics and deploy an elastically scalable infrastructure, Cloud technology is used to harvest maximum benefit of big data. The features of big data clouds are as follows

Large-scale distributed computing: A wide range of computing facilities meant for distributed systems.

Data storage: It includes all the scalable storage repositories and data services which are seamlessly available.

Metadata-based data access: The information regarding data is used instead of path and filenames.

Distributed virtual file system: Files are stored in hierarchical structure, where the nodes represent directories.

High-performance data and computation: The data as well as computation is performance driven.

Multi-dimension data: It provides support for various data types and tools for processing.

Analytics service provider: It helps to develop, deploy, and use analytics.

High availability of data and computing: The replication mechanisms for both data and computing are done so as to make them available throughout.

Table 3 Types of clouds

Sr. No	Cloud type	Description
1	Public big data clouds	The resources are made available in pay-as-go computing model and span over a large-scale industry. Such clouds are scalable elastically in terms of infrastructure. For example: Amazon big data computing in clouds [37] Windows Azure HDInsight [38] RackSpace Cloudera Hadoop [39, 40] Google cloud platform of big data computing [35]
2	Private big data clouds	The scope of these clouds are available within an enterprise which provides privacy and greater resource control through virtualized infrastructure
3	Hybrid big data clouds	An alliance of private and public big data clouds meant to achieve high availability, data sovereignty, scalability, disaster recovery and compliance, workload portability, network configuration and latency etc. During peak workloads, this deployment facilitates migration of private tasks to the public infrastructure
4	Big data access networks and computing platform	This is an integrated platform provided for data, computing and analytics where the services are delivered by multiple distinct providers

Integrated platform: It provides a platform suitable for quick analytics and deploys an elastically scalable infrastructure pending on the services provided to the end user.

The Clouds are classified into four different types on the basis of the usage, as shown below in Table 3.

Merger of big data and cloud technology give rise to many elements and services needed, which are discussed in the following texts:

Cloud Mounting Services or Mount Cloud Storage

It provides a reliable system to mount data to different cloud storages and web servers. For example: Amazon, Google drive.

Big Data Clouds

It is a platform developed for management of various data sources which includes data access, data management, programming models, schedulers, security, and so on. The big data cloud platform uses different tools like streaming, web services, and APIs for retrieving data from other sources like relational data stores, social networking, Google data, and so on.

Cloud Streaming

It is the process which transfers the multimedia data like audio and video in such a way that they are available steady and in continuous manner over social media platform.

Social Media

FaceBook, LinkedIn, Twitter, YouTube.

HTTP, REST Services

These services are used to create APIs for web-based applications which is light weight, provides high scalable and maintainable. For example: Google Earth, USGS Geo Cloud.

Big Data Computing Platform

This platform provides all those modules which are required to manage various data sources such as Compute and Data aware scheduling, Data-intensive programming models, Data security, Analytics, Distributed file system, Appliances. Data scientist is an analytic and a developer who has an access to the computing platform.

Computing Cloud

This section provides required computing infrastructure such as Virtual/Physical compute and storage from private/public/hybrid clouds.

4.1 Big Data Cloud Reference Architecture

The four-layered big data cloud reference architecture is shown in Fig. 4. The cloud infrastructure for big data should be efficient enough to handle extensive storage, high scalability, complex computing, and networking infrastructure. The infrastructure as a service providers mainly deal with servers, networks as well as storage applications and offers facilities including virtualization, basic monitoring and security, operating system, server at data center and storage services. The four layers of big data cloud architecture are discussed below:

Big Data Analytics-Software as a Service (BDA-SaaS (Analytics applications))

Big data analytics offered as service provides users with a capability to quickly work on analytics without investing on infrastructure and pay only for the resources consumed. The functions performed by this layer are

- Organization of software applications repository
- Deployment of software applications on the infrastructure
- Delivery of end results to the users.

Big Data Analytics-Platform as a Service (BPaaS (Platform for data, API's, Scheduling, Analytics, Programming environment))

This is the second layer of the architecture and considered as core layer which provides platform related services to work with stored big data and perform computing. Data management tools, schedulers, and programming environment for data-intensive data processing, which are considered as middleware management tools

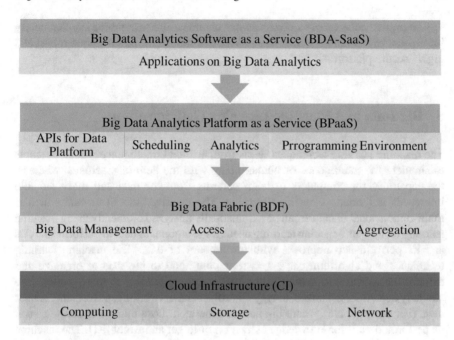

Fig. 4 Architecture of big data

resides in this layer. This layer is responsible to develop tools and software development kits which are essential component for the design of analytics.

Big Data Fabric (BDF-(Data Management, Access and aggregation))

Big data fabric layer is responsible to address tools and APIs to support data storage, computation of data and access to various application services. This layer consists of interoperable protocol and APIs responsible to connect multiple cloud infrastructure standards specified [89, 90].

Cloud Infrastructure (CI-(Computing, Storage, Network))

As discussed earlier the infrastructure handles extensive storage, high scalability, complex computing and networking infrastructure. The cloud infrastructure efficiently handles the infrastructure for storage and computation as services. The services offered by CI layer are as follows:

- To set up big data large-scale elastic infrastructure to be deployed on demand
- Creation of virtual machines dynamically
- On-demand creation of storages pertaining to large-scale data management for file/block/object-based
- To help data passage seamlessly across the storage repositories
- To create an ability to the make virtual machines and automatically mount the file system with the compute node.

The implementation of cloud can enable organizations to combine supplementary infrastructure technologies such as software defined parameters to create robust and highly secure platforms.

5   Big Data and Machine Learning Collaboration

The machine learning techniques stems from artificial intelligence, which attempts to simulate the intelligence of human being with the help of machines; whereas, the capability for computers to learn patterns from the data and make suitable inferences is termed as machine learning. The massive amount of data which is multi-dimensional in nature can be automatically analyzed efficiently using machine learning technique. Machine learning needs large amount to data to perform analysis and its performance improves with the amount of data. The machine learning techniques and algorithm can act as a helping hand to big data to organize and extracts the valued information. It has numerous applications in many sectors like anomaly detection, genomics, energy control, social networking, advertising, sensor data, risk analysis, etc., where big data is generated. Data mining or analytics can be performed on it for knowledge extraction or to get an insight [91]. The machine learning algorithms can be classified along many different perspectives.

5.1   Machine Learning Algorithm Taxonomy

The objective of machine learning is to train computers to be capable to accomplish tasks without the intervention of any human being. Eventually, machine learning algorithms involve models for identifying and validating big data to optimize a performance criterion of a process, customer, or organization. Machine learning algorithms are basically categorized into supervised, unsupervised, semi-supervised, and re-enforcement learning as shown in Fig. 5. One category of algorithm is designed for prediction (supervised learning), whereas other category is designed for observation (unsupervised learning). Hence, learning is not only a question of remembering but also of generalization to unseen cases. All the categories are discussed in detail in the following text.

Supervised Learning

The first category of machine learning algorithm is supervised learning which maps input data to a given target value also known as class label(s). Supervised learning is synonymous of classification where algorithm is trained for every input with corresponding target, which after sufficient training will be able to provide target for any new input. To achieve this, the supervised learning algorithm iteratively makes predictions on the training data and is validated before use. Learning halts when

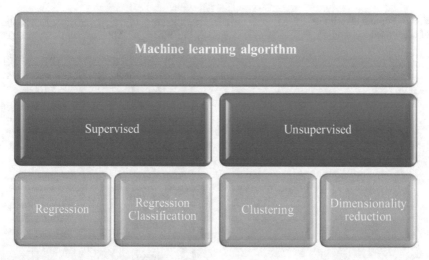

Fig. 5 Taxonomy of machine learning

the acceptable level of performance is achieved by the algorithm. The algorithm is further categorized into classification and regression type algorithms.

Classification: The algorithm predicts categorical labels called as classes. For example: malignant and benign, male and female etc.
Regression/Prediction: The algorithm predicts continuous-valued variables. For example height, weight, cost, etc.

Time series prediction and recommendations are some common types of problems built on top of classification and regression. Some popular examples of supervised machine learning algorithms for classification and regression are shown in Fig. 6.

Unsupervised Learning

The next category of machine learning algorithm is unsupervised learning, which does not consider class or labels but learns the hidden pattern. Unsupervised learning is synonymous for clustering. The objective of unsupervised learning is to discover regularities and irregularities in a set of observations, and present in the form of pattern. The process is known as density estimation in statistics is broken down into two categories: discovery of data clusters and discovery of latent factors. The methodology consists of processing input data to understand patterns similar to the natural learning process in infants or animals. Unsupervised learning does not require labeled data; however it can be used to discover classes from data. Since training data are not labeled, the learnt model cannot explain the semantic meaning of the cluster found.

The algorithm is further categorized into clustering and source signal separation. The necessary condition for the input data is the presence of representative features that can be exploited for meaningful knowledge discovery. This technique is especially suited to big data as applications have easy access to an abundance of large

Fig. 6 Popular algorithms
of supervised learning

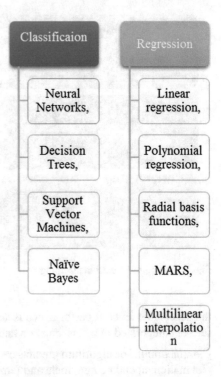

unlabeled datasets that can be processed with this learning framework. Some of the
most widely used tools in this category are K-Means Clustering, Gaussian Mixture Modeling, Spectral Clustering, Hierarchical Clustering, Principal Component
Analysis, Independent Component Analysis.

Reinforcement Learning

When a mapping function is used to learn between observations and actions with an
aim to maximize reward function, it is called as reinforcement learning algorithm
type. To maximize reward over a certain time periods, the algorithm optimized to
take a necessary action. These kinds of algorithms are more popular in robotics, but
they are being customized as per the requirements of business intelligence data applications. The tools used for reinforcement Learning are Markov Decision Process,
Q-Learning.

Semi-Supervised Classification

This algorithm type combines the functions of both supervised and unsupervised
learning. This is done by using small amounts of labeled data and which is further
mixed with large and unlabeled datasets so that it can be approximated as an appropriate learning algorithm. Usually, labeled category is used to learn class models
The big data analysis makes use of such large unlabeled datasets as oppose to traditional supervised learning algorithms, which does not caters the data requirement.
The Fig. 7 shows the list of algorithms belonging to semi-supervised classification.

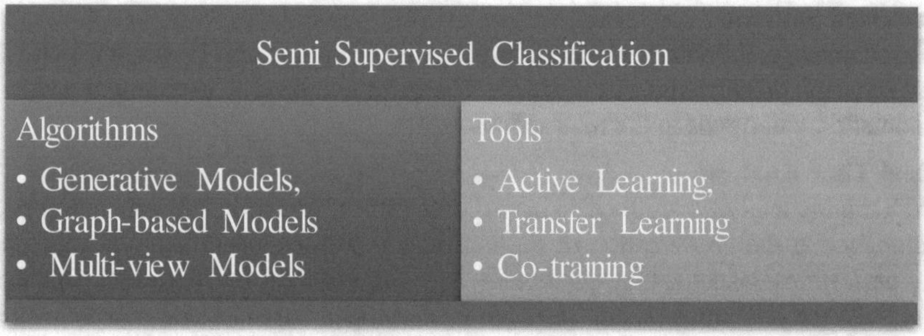

Fig. 7 Algorithms and tools of semi-supervised classification

Machine learning techniques have been found very effective and relevant to many real-world applications in health care, bioinformatics, network security, banking and finance and transportations. Over the period of time, data is created and accumulated continuously in each vertical, resulting in an incredible volume of data and known as big data. Therefore to handle big data, the machine learning methods should essentially have characteristics like robust with high velocity, transparent to variety, incremental and distributed. Machine learning has been the most utilized tool for data analytics and its performance improves with volume, variety, velocity and veracity, to give maximum value of big data.

6 Big Data: Challenges and Opportunities

Big data is generated in a rapid speed and its availability is treated as a new asset. In the current scenario, almost all major players in software and service sector, in some way has made a plunge into the big data movement and adopted big data analytics, as it appears as an evolving and significant market. However, the following discussion underlines challenges as well as opportunities associated with big data's big name.

6.1 Challenges with Big Data and Big Data Analytics

Data tornado/torrent
First, the user-generated contents such as audio, video, text, tweets, social media posts created by worldwide user using multiple accounts are more than 175 millions [92].

Assorted Sources

Next challenge, the data producers of big data are heterogeneous in nature which asks for security, privacy, and fast retrieval. The extra overhead is to examine patterns, anomalies, and trends in the transactions.

Real-Time Analysis

Web mining analyzes online data in real time or near real time that help enterprises in analyzing the conversation of the customers, perform sentiment analysis, etc., by performing visualization, prediction, and optimization. For such real-time data analysis, the predictive models needs to be developed based on real time as well as historical data. For example, with the information about wages, medical claims, and voice recordings, there is a better chance of identifying suspected fraudulent claims in the early stages.

Cyber-Attacks

Huge investments made by the government in authorized access management is driving the market for cloud identity and access management (IAM) services in this sector. Increasing number of cyber-attacks, lack of centralized access management, and data breaches would create numerous opportunities for the manufacturers to come up with more secure identity and access management services.

Market Competition

The increasing demand of cloud-based services are creating more opportunities for the companies, enabling it to build more security solutions for customers. IBM provides various security services, which are in demand in the market and creating more growth opportunities. However, tough competition in the cloud computing market is posing some hurdles for the company's growth. The profit margin in this market is attracting new players, which is eating into the company's market share.

Frauds: Key Market Players

The market is at its growing stage and is expected to rise exponentially due to enhancement of technology and increasing fraudulent activities. Fraud using sophisticated technology enables techniques, is being witnessed in every part of the globe. Fraud analytic solutions help in analyzing variety of forms of stored data and gather fraudulent information and convert the same to actionable insights, which thereby helps fraud prevention. The key market players are developing analytical solution to combat fraud. For instance, in 2016 IBM Corporation acquired IRIS Analytics, a privately held company specializing in real-time analytics to help combat fraud. This acquisition will provide technology and consultancy services to beat electronic payment fraud.

Frauds: Small Businesses

Small businesses are more vulnerable to the fraudulent activities due to lack of risk awareness. Additionally, SMEs cannot manage to pay for the layered defense security strategy, which is generally present in large enterprises. Therefore, the need to implement cost-effective security solutions, such as encryption, risk compliance, and incident management drives the adoption of fraud detection and prevention solutions. Similarly, according to Association of Certified Fraud Examiners (ACFE) one-fourth of all small businesses are vulnerable to frauds that involve check tampering. This in turn boosts the SME user segment to implement security defense measures, such as antivirus software, malware and spyware detection software and use dedicate computer for banking to combat against frauds.

6.2 Big Data: A New World of Opportunities

Big data provides big opportunities for those businessmen, technocrats who have sufficient infrastructure for implementing their big data projects. In the era of digitalization, the competitive advantage is by blend of technological developments in naïve and mature technology. The technology such as artificial intelligence, cognitive computing, cloud computing, Internet of Things, machine learning and automation will be in supporting role to amplify the benefits of big data. Big data has created a wave of innovation in many sectors and still many sectors are being yet to unleash opportunities of big data. The businesses need to be opportunistic so as to promote growth of organization in sectors such as social networking, health care, business, retail, etc. Every single business transaction should be implemented using the crux of big data technology.

7 Conclusion

Big data is available everywhere in the all the domains. Its deluge poses us with the urgency to have the mechanism and techniques in place for data acquisition, data storage and data analytics. In this chapter, we have tried to touch upon the all the aspects of big data as well as relevance of big data with cloud technology and machine learning. There is a drastic change in the way the data is dealt in traditional methods as oppose to big data. In Sect. 2, we have discussed the work which has already been done in the past for the benefit of our readers. Further, the history of big data is discussed from its inception to the current market state, along with the various dimensions of big data and its taxonomy. The lifecycle phases and the various transformations in the data are explained further.

The upward tendency of businesses to implement cloud technology and also the availability of big data is the basis of Sect. 4, which explains the need of cloud computing with the growth of big data. This section also throws light on big data cloud reference architecture to apprehend the characteristics of the market. The increasing demand of cloud-based services has created lucrative opportunities providing a wide catalog of security services, thereby driving the company's growth. However, stiff competition and advent of new entrants have restricted the growth of the company. Big data analytics with cloud computing is a technical leader in the field of data management, access, aggregation, analytics and security solutions providing constant introduction of technologically advanced security products. In addition, strong infrastructure and market factors are expected to drive the growth of the company. Introduction of new products and services would create ample growth opportunities for the businesses to prosper across globe in future.

The next section of the chapter presented the various machine learning algorithms that can be used with big data to give a big boost to big data analytics. The section briefly discussed supervised learning, unsupervised learning, semi-supervised learning and reinforcement learning. Finally, the challenges and opportunities related big data is presented for admirers and critics. It is clearly evident that future lies in big data and analytics, hence fundamental research for big data must be encouraged so that we are ready and equipped to reap maximum benefit of big data, which is referred as new soil by journalist and designer David McCandless.

References

1. Minelli, M., Chambers, M., Dhiraj, A.: Big Data Analytics. Wiley CIO Series (2014)
2. http://strata.oreilly.com/2010/01/roger-magoulas-on-big-data.html
3. McKinsey Global Institute: Big Data: The Next Frontier for Innovation, Competition and Productivity, June 2011
4. Chen, C.L.P., Zhang, C.Y.: Data-intensive applications, challenges, techniques and technologies: a survey on big data. Inform. Sci. https://doi.org/10.1016/j.ins.2014.01.015
5. Chen, M., Mao, S., Liu, Y.: Big data survey. Mob. Netw. Appl. **19**(2), 171–209 (2014)
6. Chen, J., Chen, Y., Du, X., Li, C., Lu, J., Zhao, S., Zhou, X.: Big data challenge: a data management perspective. Front. Comput. Sci. **7**(2), 157–164 (2013)
7. Wu, X., Zhu, X., Wu, G.Q., Ding, W.: Data mining with big data. IEEE Trans. Knowl. Data Eng. **26**(1), 97–107 (2014)
8. Kaisler, S., Armour, F., Espinosa, J.A, Money, W.: Big data: issues and challenges moving forward. In: Proceedings of the 46th IEEE Annual Hawaii international Conference on System Sciences (HICC 2013), Grand Wailea, Maui, Hawaii, pp. 995–1004, Jan 2013
9. Assuncao, M.D., Calheiros, R.N., Bianchi, S., Netto, M., Buyya, R.: Big data computing and clouds: trends and future directions. J. Parallel Distrib. Comput. (JPDC) **79**(5):3–15 (2015)
10. Survey of big data architectures and framework from the industry, NIST big data public working group (2014). http://jtc1bigdatasg.nist.gov/_workshop2/07_NBD-PWD_Big_Data_Architectures_Survey.pdf. Last accessed 30 Apr 2014
11. Mayer, V.V., Cukier, K.: Big Data: A Revolution That Will Transform How We Live, Work and Think. John MurrayPress, UK (2013)
12. Team, O.R.: Big Data Now: Current Perspectives from O'Reilly Radar. O'Reilly Media Sebastopol, CA, USA (2011)
13. Gantz, J., Reinsel, D.: Extracting value from chaos. In: Proceedings of the IDC iView, pp. 1–12 (2011)
14. Zikopoulos, P., Eaton, C.: Understanding Big Data: Analytics for Enterprise Class Hadoop and Streaming Data. McGraw-Hill, New York, NY, USA (2011)
15. Meijer, E. Theworld according to LINQ. Commun. ACM **54**(10), 45–51 (2011)
16. Laney, D.: 3d data management: controlling data volume, velocity and variety. Gartner, Stamford, CT, USA, White Paper (2001)
17. Manyika, J., et al.: Big data: The Next Frontier for Innovation, Competition, and Productivity, pp. 1–137. McKinsey Global Institute, San Francisco, CA, USA (2011)
18. Cooper, M., Mell, P.: Tackling Big Data (2012). http://csrc.nist.gov/groups/SMA/forum/documents/june2012presentations/f%csm_june2012_cooper_mell.pdf

19. Brewer, E.A.: Towards robust distributed systems, keynote speech. In: 19th ACM Symposium on Principles of Distributed Computing (PODC 2000), Portland, Oregon, July 2000
20. Gray, J.: The transaction concept: virtues and limitations. In: Proceedings of the 7th International Conference on Very Large Databases (VLDB' 81), vol. 7, pp. 144–154 (1981)
21. Pritchett, D.: BASE: an ACID alternative. Queue Object Relat. Mapp. **6**(3), 48–55 (2008)
22. Borkar, V.R., Carey, M.J., Li, C.: Big data platforms: what's next? XRDS, Crossroads, ACM Mag. Students, vol. 19, no. 1, pp. 44_49 (2012)
23. Borkar, V., Carey, M.J., Li, C.: Inside big data management: Ogres, onions, or parfaits? In: Proceedings of the 15th International Conference Extending Database Technology, pp. 3–14 (2012)
24. Dewitt, D., Gray, J.: Parallel database systems: the future of high performance database systems. Commun. ACM **35**(6), 85–98 (1992)
25. Teradata. Teradata, Dayton, OH, USA (2014). http://www.teradata.com/
26. Netezza. Netezza, Marlborough, MA, USA (2013). http://www-01.ibm.com/software/data/netezza
27. Aster Data. ADATA, Beijing, China (2013). http://www.asterdata.com/
28. Greenplum. Greenplum, San Mateo, CA, USA (2013). http://www.greenplum.com/
29. Vertica. http://www.vertica.com/ (2013)
30. Hey, T., Tansley, S., Tolle, K.: The Fourth Paradigm: Data-Intensive Scientific Discovery. Microsoft Res, Cambridge, MA, USA (2009)
31. Ghemawat, S., Gobioff, H., Leung, S.-T.: The Google file system. In: Proceedings of the 19th ACM Symposium Operating Systems Principles, pp. 29–43 (2003)
32. Dean, J., Ghemawat, S.: Mapreduce: simplified data processing on large clusters. Commun. ACM **51**(1), 107–113 (2008)
33. Noguchi, Y.: The Search for Analysts to Make Sense of Big Data, National Public Radio, Washington, DC, USA (2011). http://www.npr.org/2011/11/30/142893065/the-search-foranalysts-to-make-%sense-of-big-data
34. Apache Spark. https://spark.incubator.apache.org. Last accessed 03 Apr 2014
35. Google big query. https://cloud.google.com/bigquery-tour. Last accessed 15 Jan 2015
36. Chang, F., Dean, J., Ghemawat, S., Heish, W.C., Wallach, D.A., Burrows, M., Chandra, T., Fikes, A., Gruber, R.E.: Bigtable: a distributed storage system for structured data. In: Proceedings of the 7th USENIX Symposium on Operating Systems Design and Implementation (OSDI 2006), Seattle, WA, Nov 2006
37. Amazon elastic MapReduce, developer guide (2015). http://docs.aws.amazon.com/ElasticMapReduce/latest/DeveloperGuide/emr-dg.pdf. Last accessed 1 Nov 2014
38. Chauhan, A., Fontama, V., Hart, M., Hyong, W., Woody, B.: Introducing Microsoft Azure HDInsight, Technical Overview. Microsoft press, One Microsoft Way, Redmond, Washington (2014)
39. Rack space. www.rackspace.com. Last accessed 22 Aug 2014
40. Horton Hadoop. http://hortonworks.com. Last accessed 22 Aug 2014
41. Cloudera Hadoop. http://www.cloudera.com. Last accessed 03 Sep 2014
42. Buyya, R., Vecchiola, C., Selvi, T.: Mastering in Cloud Computing—Foundations and Applications Programming. Morgan Kaufman, USA (2013)
43. Decandia, G., Hastorun, D., Jampani, M., Kakulapati, G., Lakshman, A., Pilchin, A., Sivasubramanian, S., Vosshall, P., Vogels, W.: Dynamo: Amazon's highly available key-value store. In: Proceedings of the 21st ACM Symposium on Operating Systems Principles (SOSP 2007), Stevenson, Washington, USA, Oct 2007
44. Oracle Berkeley DB, Oracle data sheet. http://www.oracle.com/technetwork/products/berkeleydb/berkeley-dbdatasheet-132390.pdf. Last accessed 03 Sep 2014
45. MongoDB operations best practices. http://info.10gen.com/rs/10gen/images/10gen-/mongoDB_Operations_Best_Practices.pdf
46. Apache couch DB, a database for the web. www.couchdb.apache.org. Last accessed 10 Sep 2014
47. United Nations Global Pulse: Big Data for Development: A Primer (2013)

48. Gantz, J., Reinsel. D.: The digital universe in 2020: big data, bigger digital shadows, and biggest growth in the far east. In: Proceedings of the IDC iView, IDC Analyze the Future (2012)
49. Apache MapReduce. http://hadoop.apache.org/docs/stable/mapred_tutorial.html. Last accessed 20 Feb 2015
50. Buyya, R., Shin Yeo, C., Venugopal, S., Brobergand, J., Brandic, I.: Cloud computing and emerging IT platforms: vision, hype, and reality for delivering computing as the 5th utility. Fut. Gener. Comput. Syst. **25**(6), 599–616 (2009)
51. The anatomy of big data computing Raghavendra Kune,*,†, Pramod Kumar Konugurthi, Arun Agarwal, Raghavendra Rao Chillarige and Rajkumar Buyya
52. Cisco Syst., Inc.: Cisco visual networking index: global mobile data traffic forecast update. Cisco Systems, Inc., San Jose, CA, USA, Cisco Technical Report 2012-2017, 2013
53. Gallagher, F.: The Big Data Value Chain (2013). http://fraysen.blogspot.sg/2012/06/big-data-value-chain.html
54. Sevilla, M.: Big Data Vendors and Technologies, the list! (2012). http://www.capgemini.com/blog/capping-it-off/2012/09/big-data-vendors-a%nd-technologies-the-list
55. What is Big Data, IBM, New York, NY, USA (2013). http://www-01.ibm.com/software/data/bigdata/
56. Evans, D., Hutley, R.: The explosion of data. In: White Paper (2010)
57. KnowWPC: eBay Study: How to Build Trust and Improve the Shopping Experience (2013). http://knowwpcarey.com/article.cfm?aid=1171
58. Gantz, J., Reinsel, D.: The digital universe decade-are you ready. In: Proceedings of White Paper, IDC (2010)
59. Layton, J.: How Amazon Works (2013). http://knowwpcarey.com/article.cfm?aid=1171
60. Cukier, K.: Data, data everywhere. In: Economist, vol. 394, no. 8671, pp. 3–16 (2010)
61. Bryant, R.E.: Data-intensive scalable computing for scientific applications. Comput. Sci. Eng. **13**(6), 25–33 (2011)
62. SDSS (2013). http://www.sdss.org/
63. Atlas (2013). http://atlasexperiment.org/
64. Wang, X.: Semantically-aware data discovery and placement in collaborative computing environments. Ph.D. Dissertation, Dept. Comput. Sci., Taiyuan Univ. Technol., Shanxi, China, 2012
65. Middleton, S.E., Sabeur, Z.A., Löwe, P., Hammitzsch, M., Tavakoli, S., Poslad, S.: Multi-disciplinary approaches to intelligently sharing large volumes of real-time sensor data during natural disasters. Data Sci. J. **12**, WDS109–WDS113 (2013)
66. Selavo, L., et al.: Luster: wireless sensor network for environmental research. In: Proceedings of the 5th International Conference Embedded Networked Sensor Systems, pp. 103–116, Nov 2007
67. Barrenetxea, G., Ingelrest, F., Schaefer, G., Vetterli, M., Couach, O., Parlange, M.: Sensorscope: out-of-the-box environmental monitoring. In: Proceedings of the IEEE International Conference Information Processing in Sensor Networks (IPSN), pp. 332–343 (2008)
68. Wahab, M.H.A., Mohd, M.N.H., Hanafi, H.F., Mohsin, M.F.M.: Data pre-processing on web server logs for generalized association rules mining algorithm. In: World Academy Science, Engineering Technology, vol. 48, p. 970 (2008)
69. Nanopoulos, A., Manolopoulos, Y., Zakrzewicz, M., Morzy, T.: Indexing web access-logs for pattern queries. In: Proceedings of the 4th International Workshop Web Information Data Management, pp. 63–68 (2002)
70. Joshi, K.P., Joshi, A., Yesha, Y.: On using a warehouse to analyze web logs. Distrib. Parallel Databases **13**(2), 161–180 (2003)
71. Laurila, J.K., et al.: The mobile data challenge: big data for mobile computing research. In: Proceedings of the 10th International Conference Pervasive Computing, Workshop Nokia Mobile Data Challenge, Conjunction, pp. 1–8 (2012)
72. Castillo, C.: Effective web crawling. In: ACM SIGIR Forum, vol. 39, no. 1, pp. 55–56 (2005)
73. Choudhary, S., et al.: Crawling rich internet applications: the state of the art. In: Proceedings of the Conference of the Center for Advanced Studies on Collaborative Research, CASCON, pp. 146–160 (2012)

74. Ghani, N., Dixit, S., Wang, T.-S.: On IP-over-WDM integration. IEEE Commun. Mag. **38**(3), 72–84 (2000)
75. Manchester, J., Anderson, J., Doshi, B., Dravida, S.: Ip over SONET. IEEE Commun. Mag. **36**(5), 136–142 (1998)
76. Farrington, N., et al.: Helios: a hybrid electrical/optical switch architecture for modular data centers. In: Proceedings of the ACM SIGCOMM Conference, pp. 339–350 (2010)
77. Wang, G., et al.: C-through: part-time optics in data centers. SIGCOMM Comput. Commun. Rev. **41**(4), 327–338 (2010)
78. Ye, X., Yin, Y., Yoo, S.B., Mejia, P., Proietti, R., Akella, V.: DOS_A scalable optical switch for datacenters. In: Proceedings of the 6th ACM/IEEE Symposium on Architectures for Networking and Communications Systems, pp. 1–12, Oct 2010
79. Singla, A., Singh, A., Ramachandran, K., Xu, L., Zhang, Y.: Proteus: a topology malleable data center network. In: Proceedings of the 9th ACM SIGCOMM Workshop Hot Topics in Networks, pp. 801–806 (2010)
80. Liboiron-Ladouceur,O., Cerutti, I., Raponi, P.G., Andriolli, N., Castoldi, P.: Energy-efficient design of a scalable optical multiplane interconnection architecture. IEEE J. Sel. Topics Quantum Electron. **17**(2), 377–383 (2011)
81. Kodi, K., Louri, A.: Energy-efficient and bandwidth-reconfigurable photonic networks for high-performance computing (HPC) systems. IEEE J. Sel. Topics Quantum Electron. **17**(2), 384–395 (2011)
82. Müller, H., Freytag, J.-C.: Problems, methods, and challenges in comprehensive data cleansing. Professoren des Inst. Für Informatik (2005). http://www.dbis.informatik.hu-berlin.de/_leadmin/research/papers/techreports/2003-hubib164-mueller.pdf
83. Noy, N.F.: Semantic integration: a survey of ontology-based approaches. In: ACM Sigmod Record, vol. 33, no. 4, pp. 65–70 (2004)
84. Lenzerini, M.: Data integration: a theoretical perspective. In: Proceedings of the 21st ACM SIGMOD-SIGACT-SIGART Symposium Principles Database Systems, pp. 233–246 (2002)
85. Maletic, J.I., Marcus, A.: Data cleansing: beyond integrity analysis. In: Proceedings of the Conference on Information Quality, pp. 200–209 (2000)
86. Zhang, Y., Callan, J., Minka, T.: Novelty and redundancy detection in adaptive filtering. In: Proceedings of the 25th Annual International ACM SIGIR Conference on Research and Development in Information Retrieval, pp. 81–88 (2002)
87. Salomon, D.: Data Compression. Springer, New York, NY, USA (2004)
88. Tsai, T.-H., Lin, C.-Y.: Exploring contextual redundancy in improving object-based video coding for video sensor networks surveillance. IEEE Trans. Multimed. **14**(3), 669–682 (2012)
89. Baah, G.K., Gray, A., Harrold, M.J.: On-line anomaly detection ofdeployed software: a statistical machine learning approach. In: Proceedings of the 3rd International Workshop Software Quality Assurance, pp. 70–77 (2006)
90. Moeng, M., Melhem, R.: Applying statistical machine learning tomulticore voltage and frequency scaling. In: Proceedings of the 7th ACM International Conference on Comput. Frontiers, pp. 277–286 (2010)
91. Han, J., Kamber, M., Pei, J.: Data mining: concepts and techniques. Elsevier, Burlington, MA (2012)
92. Kelly, J.: Taming Big Data (2013). http://wikibon.org/blog/taming-big-data/

Cloud Computing Based Knowledge Mapping Between Existing and Possible Academic Innovations—*An Indian Techno-Educational Context*

P. K. Paul, Vijender Kumar Solanki and P. S. Aithal

Abstract Cloud Computing is a revolutionary name in today's age and not only in Information Science and Technology (IST). This is a type of virtualization of the hardware, software, and systems. Cloud Computing allows healthy and wider efficient computing services in terms of providing centralized services of storage, applications, operating systems, processing, and bandwidth. Cloud Computing is a type of architecture which helps in promotion of scalable computing. Cloud Computing is also a kind of resource sharing platform and thus needed in almost all the spectrum and areas regardless of its type. Today Cloud Computing has a wider market and it is growing rapidly. The manpower in this field is mainly outsourced from the IT and Computing, but this is an urgent need to offer Cloud Computing as a full-fledged Bachelors and Masters programs. In India also Cloud Computing is rarely seen as an education program, but the situation is now changing. There are many potential to offer Cloud Computing in Indian educational segment. This paper is conceptual in nature and deals with the basic of Cloud Computing; its need, features, types existing, and possible programs in the Indian context. Paper also proposed several programs which ultimately may be helpful for building solid Digital India.

Keywords Cloud computing · Virtualization · Fog computing · Green systems Cloud systems · Manpower · Education · Universities · New edge programs

P. K. Paul (✉)
Raiganj University, West Bengal, India
e-mail: pkpaul.infotech@gmail.com

V. K. Solanki
CMR Institute of Technology (Autonomous), Hyderabad, India
e-mail: spesinfo@yahoo.com

P. S. Aithal
Srinivas University, Mangaluru, Karnataka, India
e-mail: psaithal@gmail.com

© Springer Nature Singapore Pte Ltd. 2019
M. Mittal et al. (eds.), *Big Data Processing Using Spark in Cloud*,
Studies in Big Data 43, https://doi.org/10.1007/978-981-13-0550-4_4

1   Introduction

The fundamental of Cloud Computing is the approach and model for building virtualization of many computing and information technology products and services. Cloud Computing mainly depends on the internet and network technology [1, 2]. The enhancement of grid computing into more scalability is the main task and theme of the Cloud Computing. The software, hardware, application, operating systems may be possible to get by the cloud implementation into traditional computing systems. Today organizations, institutions, government bodies, companies are moving towards Cloud Computing services. Though there are delay and shortage in ready manpower [3–5]. Many universities around the world have started several programs on cloud computing and similar areas. While in the Indian context, the Cloud Computing education is still dry. Though universities, colleges need to start programs on Cloud Computing for healthy and solid manpower development and healthy services.

1.1   Aim and Agenda

The main aim and agenda of this research work include but not limited to the following basic mentioned:

- To know basic about the Cloud Computing and Virtualization as a tool and its solid role in the industry, academia, health care, etc.
- To learn about the Cloud Computing along with its basic features and characteristics in a simple context.
- To find out the types of Cloud Computing and its service offering with models in a brief manner.
- To learn about the Indian education systems and its contemporary situation with succinct details and so on.
- To dig out the Cloud Computing and Virtualization and its market in International and Indian context.
- To learn about the healthy information in Cloud Computing educational programs including the research programs available in this field.
- To draw the healthy Cloud Computing educational and research programs in Indian educational context.

1.2   Methodology Undertaken

For this study and conceptual research work various tools, as well as techniques, are used and out of which few important are the review of the literature. Though, the web review also plays a sophisticated role to find out latest on Cloud Computing and running courses in the field. To find out the educational situation in India, several sites have been analyzed such as UGC; Government of India, AICTE; Government of India, etc. More than 700 educational institutes are assessed during this study including the Universities (*public and private*), Deemed Universities, and Central Universities of Indian territories. The Review of related Journals and major findings of the events such as conferences, etc., have been analyzed to know latest about Cloud Computing including its issues especially related to the Education, Training, Research in Cloud Computing and Engineering.

1.3 Hypothesis and Assumption

The study is conceptual-based and deals with the techno-management studies with a focus on educational strategies and management. The core and main hypothesis of this work are not limited to the

- Cloud Computing and virtualization is an important domain of Applied Science and deals with the several interdisciplinary facets.
- Cloud Computing has evolved and integrates new domains and concepts such as Green Computing, Green Information Systems, Environmental Informatics, Energy Informatics, Green Information Sciences, etc.
- Cloud Computing is available with many corporate vendors as a specialization in their networking and similar certifications.
- Cloud Computing specializations and full-fledged degrees are rising around the world and with several levels of study.
- In India Cloud Computing programs mainly available as educational modules, chapters, etc., rather as a university or IIT based programs.
- Cloud Computing has the possibilities as engineering, science, and management programs in the Indian context.

2 Cloud Computing and Virtualization: Types, Services, and Emerging Interdisciplinary Domains

Cloud Computing is the architecture which is based on several virtualization principles. Most of the IT and Computing tools such as software; both application and operating systems, Hardware, IT devices positively possible to store with the help of Cloud Computing. The foundations of the Cloud Computing are cloud platform and cloud architecture [6–8]. The services normally provided by the third party. The Cloud Computing normally believes in less computing infrastructure uses and less economic involvement. The Cloud Computing basically depends on Green and Environmental Computing practices. The Cloud Computing basically depends on Green and Environment computing practices. Cloud Computing and its basic features are as follows:

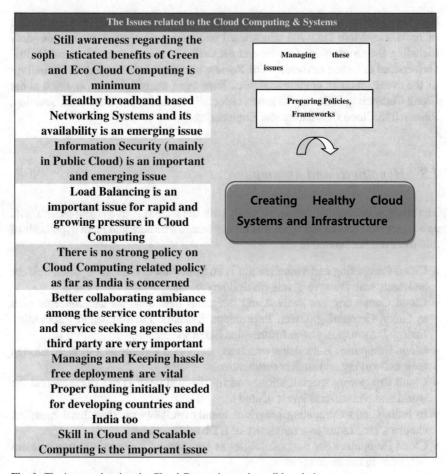

Fig. 1 The issues related to the Cloud Computing and possible solutions

- Cloud Computing normally helps to build sophisticated hardware, software along with applications round o' clock and possible to get anywhere and anytime by dedicated broadband Networks.
- It improves greater efficiencies in information activities including the reduction of Information Technology and Computing cost through the hassle free deployment of databases.
- Cloud Computing gives us easily available IT services, systems, and solutions.
- It create a Green and Energy efficient computing platform and ultimately builds Green and eco-friendly Infrastructure [9–11].

While the Cloud Computing also suffers from some of the issues and problems and these are listed in Fig. 1. Cloud Computing is classified into several types and these are mainly categorized as Public Cloud, Private Cloud, and Hybrid Cloud.

Public Cloud is available with network and opens for the consumer based on their need [3, 12]. The organization may choose their services, depending upon need. In Public cloud services, the Internet plays a vital role and it is not offered by the internet directly; though practically it provides many services which include the elasticity, cost-effective services along with easy deployment of services and the solutions. Moreover, there is no need of deploying, managing as well as securing the infrastructure [13–15].

Private Cloud Computing is the platform which is operated by the concerned organization and it is a kind of in-house services. However it may be managed internally (or managed by the third party and hosted internally) or externally. However this kind of cloud may be managed by fully third party and there is a need to perform all the activities with the direct and proper supervision of concerned company or foundations [11, 16, 17].

Hybrid Cloud Computing is the Combination of Private Cloud Computing along with Public Cloud Computing. Though, it may also blend of more than two Cloud Computing platforms; which are depicted later on. Practically it offers several advantages than that of Public Cloud or Private Cloud in many contexts. Virtually, newer and healthy implementation models have also been developed to merge models to an integrated solution [3, 18]. The Hybrid Cloud Computing is applicable for the outsourcing of non-business critical information and processing to the public cloud. Importantly, apart from these three types the Cloud Computing, it may also be classified as

- *Community Cloud Computing.*
- *Shared Private Cloud Computing.*
- *Dedicated Private Cloud Computing.*
- *Dynamic Private Cloud Computing.*
- *Controlled Cloud Computing.*
- *Distributed Cloud Computing.*

The above-mentioned aspects are deal with the types of Cloud. However, based on service models the cloud computing may be depicted as under [1, 19, 20]

- Platform-as-a-Services/PaaS (*Where platform oriented services basically offered*).
- Software-as-a-Service/SaaS (*In which Software and Systems mainly provided by the Internet*).
- Infrastructure-as-a-Service/IaaS (*Where infrastructure and software and hardware systems available through the Internet and intranet*).
- Desktop-as-a-Service/DaaS (*In which desktop applications become easy*).
- Security-as-a-Service (*Where security and assurance basically provided*).

2.1   Cloud Computing, Education Segment and Educational Opportunities: Indian Context

The research revealed that up to 1999, Indian software market was just 1.7 billion US dollar. And after 10 years it has touched about 12.5 billion dollar. Regarding the exports, the amount reaches around 4 billion (*in USD*) in 1999. Compared amount was 150 million (*in US dollar*) in 1990–1999.

Today many developed countries are offering cloud computing services, but the developing countries are also having the potentiality to offer the Cloud and similar services. Initially it may be started inside the territory (*or within the country*) with huge potentiality to establish the business from the outside world due to healthy relationship (regarding IT) with other countries like USA, UK, China, Mexico, Ireland, Russia, Malaysia, and others. Thus it is a fact that all the platforms like the private cloud, public cloud, hybrid cloud with several services may be offered. It is a fact that India is a global leader in IT segment and top five companies out of ten are from India. Thus, India has the potential to offer Cloud Computing Services. It is an important fact that many small and mid-level (*many of them are new*) companies already have started cloud and virtualization companies, the Silicon Valley [21] has listed 25 best companies as given in Table 1.

Importantly India's Internet penetration trails behind China, New Zealand, Australia, and some other major developed nations. The number of Internet users in India is rising rapidly; though it has currently few important issues. The bandwidth cost in India is also higher and far high than USA. Today India has 45 million small and medium enterprises and rapidly increasing Cloud Computing has healthy benefit and wider possibilities and thus there is a big market waiting (Table 2) for in the entire segment such as health, education, government, business, tourism, etc. According to an important statistics among the companies, IT and IT Enable Services are most popular as well as contributed and include (their IT budget for Cloud)—20–25%; compare to 10–15%, 5–8%, 2–5% in manufacturing, health care, real contributed around respectively.

According to the report of the Allied Market Research Cloud Computing (CC) application rising among all the sectors such as healthcare units, educational institutes, Business, Government marketing and supply chain management institute, etc., and also will gradually grow up and will touch around 24.1% of total sectors soon. Some of the big venture such as Google, HP, IBM are also giving efforts in Cloud and Big data along with several services, products and the salaries of such cloud professional are around 10–17+ lakhs with around 7 years experience (*The Economics Times, 2015*). In another research, Wanted Analytics, also highlighted that largest job providers in cloud segment are CISCO, Oracle and median salary is around US dollar 103,000. While in the cloud and Big data category some of the popular job titles are include Big data Scientist, data solution expert, Big Data architect, Big data cloud engineer, etc. Moreover the Cloud sales professionals and there demand in India is increasing, the Oracle India is to hire Cloud-Big data sales professional across the country. Hence the Cloud Computing field is emerging and vis-à-vis the

Table 1 Silicon valley listed 25 best companies in Cloud Computing in India

Serial no.	Company name	Establishment	Place	Services and offering
1	Alpinesoft IT Solutions	2000	New Delhi	Google Apps for business, consultancy Services and Salesforce consultancy
2	Blue Pi Consulting	2012	Gurgaon	Expertise in cloud ecosystems
3	Brio Technologies	2004	Hyderabad	Bespoke Cloud Computing solution
4	CipherCloud	2010	Hyderabad (with SJ)	Cloud visibility and security solutions
5	Clicklytics	2008	Bangalore	Cloud systems specially for hotels
6	CloudCodes	2011	Pune	Cloud-based focused for Google Apps
7	CSQ Global Solutions	2013	Chennai	All type of cloud services PaaS, SaaS, IaaS, Data analytics, mobility, digitalization
8	CTRLS	2007	Hyderabad	Private cloud on demand, cloud hosting, disaster recovery
9	Elixia Tech Solutions	2011	Mumbai	Cloud solutions with expertise in web and mobile based development
10	Expert Solution Technologies	2011	Chennai	API integration services around ERP and CRM with external system
11	IBN Technologies	1999	Pune	Cloud infrastructure and consulting services
12	Infinient	2007	Goa	Business web services to large and powerful computing instances with easy deployment
13	InteractCRM	2002	Mumbai	Cloud-based communication
14	iSynergyTechSys	2007	Pune	SaaS and cloud services focusing on ISVs
15	Mould Networks	2013	Chennai	Managed networked solutions stack and customization of hardware
16	NEC Technologies India	2005	Noida	Cloud services dedicated to network solutions

(continued)

Table 1 (continued)

Serial no.	Company name	Establishment	Place	Services and offering
17	NWDCo	2001	Mumbai	Offered several products powered by cloud platform
18	Plumsoft	2007	Hyderabad	Cloud-based ERP products, CRM systems etc.
19	SailPoint	2005	Pune	Innovative identity governance and access management solutions
20	Saturo Technologies	2014	Mumbai	SaaS, Mobile apps, Social media management
21	ShareNSearch	2012	Bangalore	Internet based social media powered by Cloud Computing
22	Sify Technologies	1995	Chennai	ICT solutions that mainly cover enterprise data center services, technology integration, infrastructure services
23	Umbrella Infocare	2013	Noida	Cloud and mobility services for core business
24	Unify	2008	Mumbai	Cloud communication software
25	Vaultize	2010	Pune	Mobile collaboration, digital right management, Mobile content management (MCM)

Table 2 Depicted Indian Cloud Computing market in USD

Indian domestic Cloud Computing market (in USD)		
Private cloud (CY 09)	250–260	In all the segment, mainly in Industries
Public cloud (CY 09)	67–93	In all the segments
Private cloud (CY 11)	160–192	In all the segment, mainly in Industries
Private cloud (CY 11)	700–720	In all the segment, mainly in Industries

job opportunities. Internationally Cloud Computing education (mainly MSc CC) is offered by some of the universities such New Castle University, Cork Institute of Technology, University of Essex, University of Leicester, Anglia Ruskin University, Staffordshire University, Middlesex University, National College of Ireland, etc. [11, 15]. Thus there are huge potentiality in Cloud Computing and related education programs such as Data Science, Distributed Computing, Grid Computing, Data Sciences, Virtualization, etc., but India is very much dry market in Cloud Computing education segment.

India is the largest education system in the world with total 40,000+ Higher Educational Institutes (HEIs). It comprises with colleges, polytechnic colleges, research centers, universities (including state, private, deemed, central, public-private, etc.), Institute of National Importance (*categorized as higher than universities*), etc. [22–24]. A brief list such institutes have depicted in Table 2. Among them many are centrally and state funded and established many years back but still Cloud Computing as a program is not all offered. However few private have started Cloud Computing programs and Table 3 depicted more on this [15, 22–24].

However, due to huge potentialities of the Cloud Computing and allied domains, the interested Indian IT professional (or future professional) may either get the Online and Distance Education in the domain as BS/MS degrees offered by many international universities and can be avail right here in India. Though there will be a missing feature on these programs, i.e., Skill development and Practical Knowledge. Importantly the certifications offered by many international IT organizations and education providers are important in this regard and many Indian Cloud Education seekers getting the benefits. Among such certification and training providers, few are Microsoft, Linux, Oracle Corporation, Cisco Systems, EC Council, Juniper Networks, ISC2, etc. A Detailed list of such Cloud Computing related programs is listed in Table 4.

Although Cloud Computing as a full-fledged degree is not at all offered in most of the universities whether central universities or state universities. Currently, there are 46 central universities and 340+ State Universities but the programs are absent ignoring its potential job values, and demand as depicted in previous section/s. Indian Education related to the Sciences comprises with *Science* and *Technology* (also referred as Engineering). In these two segments, degrees are offered as **BSc** and **B.Tech or BE** (Bachelor of Engineering) respectively. Thus in both the category Cloud Computing has the potentiality as an educational subject or specializations. Though full-fledged *either* **BSc-Cloud Computing or B.Tech/BE-Cloud Computing** are not offered in India (among world highest 40,000+ HEIs); but few deemed universities have been offering the B.Tech Specialization of Cloud Computing in the Computer Science and Engineering (CSE) and IT branches, a detail on this is depicted in Table 5. While the MSc-Cloud Computing is also absent in educational catalogs in India. Though in Masters level M.Tech is offered with full concentration in Cloud Computing or as Specialization of CSE/IT braches. All of these universities are Deemed Universities. A detail lust on this is depicted in Table 6.

Some of these programs (mainly B.Tech CSE/IT with Cloud Computing) also offered with tie-up of IBM and I-Nurture (a leading educational consultant organization and technology provider).

3   Possible Academic Programs in Indian Perspectives

As reported, India is the largest education system in the universe with huge HEIs. Many of them were tagged with University, Institute of National Importance, Research Centre/Institutes under MHRD etc., Engineering Colleges, etc. Within

Table 3 Showing Indian higher education segment and current position and future potentiality regarding Cloud Computing education

Universities and others higher educational institutions (HEIs)	The numbers	Location/states	Remarks on *Cloud Computing*
Central Universities	46	Pan India with 28 States and 5 UT	Not a single University offers Cloud Computing as a Degree/PG
State Universities	342	Pan India with 28 States and 5 UT	Not a single University offers Cloud Computing as a Degree/PG
State Private Universities	239	Except some states and 5 UT	Few have started programs with tie-up with other agencies and companies
Deemed Universities	125	Except some states and 5 UT	Few have started programs with tie-up with other agencies and companies
Indian Institute of Technology [IITs]	23	Bhubaneswar, Chennai, Delhi, Gandhinagar, Guwahati, Hyderabad, Indore, Jodhpur, Kanpur, Kharagpur, Mandi, Mumbai, Patna, Ropar, Roorkee, Varanasi, Pallakkad, Tirupati, Dhanbad, Bhilai, Goa, Jammu, Dharwad	Not a single IITs offers Cloud Computing as a Degree/PG. Though Research programs leading to PhDs are offered. And CC as a module, paper etc.
National Institute of Technology [NITs]	31	Agartala, Allahabad, Arunachal Pradesh, Andhra Pradesh, Bhopal, Calicut, Delhi, Durgapur, Goa, Puducherry, Hamirpur, Jaipur, Manipur, Meghalaya, Mizoram, Nagaland, Jalandhar, Jamshedpur, Kurukshetra, Nagpur, Patna, Raipur, Rourkela, Sikkim, Silchar, Srinagar, Surat, Karnataka, Tiruchirappalli, Uttarakhand, Warangal	Not a single NITs offers Cloud Computing as a Degree/PG. Though Research programs leading to PhDs are offered. And CC as a module, paper etc.

(continued)

Table 3 (continued)

Universities and others higher educational institutions (HEIs)	The numbers	Location/states	Remarks on *Cloud Computing*
Indian Institute of Information Technology [IIITs] *established as INI*	04	Allahabad, Jabalpur, Gwalior, Kancheepuram	Not a single IIITs offers Cloud Computing as a Degree/PG. Though Research programs leading to PhDs are offered. And CC as a module, paper etc.
Indian Institute of Management [IIMs]	19	Calcutta, Ahmedabad, Bangalore, Lucknow, Kozhikode, Indore, Shillong, Rohtak, Ranchi, Raipur, Tiruchirappalli, Udaipur, Kashipur, Nagpur, Bodh Gaya, Visakhapatnam, Amritsar, Sambalpur, Sirmaur	Not a single IIMs offers Cloud Computing or Cloud Managements as a Degree/PG. Though Research programs leading to Fellow of IIM are offered. And CC as a module, paper in Information Systems specializations
Indian Institute of Science Education and Research [IISERs]	07	Calcutta, Mohali, Thiruvanthapuram, Pune, Tirupati, Berhampur, Bhopal	Not offered, though potentialities are there to offers scientific and theoretical programs on clouds
Academy of Scientific and Innovative Research	01	Chennai (But having Research Centers Pan India basis)	Not offered, though potentialities are there to offers scientific and theoretical programs on clouds
Indian Institute of Engineering Science and Technology [IIEST], Shibpur	01	Shibpur	Not offers Cloud Computing as a Degree/PG. Though Research programs leading to PhDs are offered. And CC as a module, paper etc.
School of Planning and Architecture [SPAs]	03	Bhopal, New Delhi, Vijayawada	Not offered but has huge potentiality to offer Cloud Architecture, Cloud Designing etc. as a program specializations

Table 4 Depicted the industry offered Cloud Computing certification programs etc.

Most popular programs related to Cloud Computing	
Microsoft	Microsoft Azure/MCSE (Private Cloud) as well as others MCSE indirectly related with Cloud Computing
Linux (Red hat)	**Initial Level**: RHCA (Cloud)/RHCSA (Open stack)/RHCE (Hybrid Cloud)/RHCE (Platform-as-a-Service)
	Higher Level: RHCE (Hybrid Cloud Storage)/RHCVA/RHCE (Deployment and System Management)
Oracle Corporation	Oracle Network Virtualization/Oracle Server Virtualization/Oracle Cloud
Cisco Systems	**Initial Level**: CCNA (Cloud)/CCNA (Data Center)
	Higher Level: CCNP (Cloud)/CCNP (Data Center)/CCIE (Data Center)
EC Council	EC Council Cloud Essential
Juniper Networks	JNCDS (Data Center)/JNCDP (Data Center)
ISC2	CISSP/CCSP (Cloud Security)

Table 5 The Universities which are offering Cloud Computing in Bachelors level

Universities	Programs	Faculty/wing
Hindsthan University, TN	B.Tech IT (Cloud Computing) B.Tech CSE (Cloud Computing)	Faculty of Engineering and Technology
Graphic Era University,	B.Tech CSE (Cloud Computing)	Faculty of Engineering and Technology
University of Petroleum and Energy Studies, Uttarakhand	B.Tech CSE (Cloud Computing)	Faculty of Engineering and Technology
Sharda University, UP	B.Tech CSE (Cloud Computing)	Faculty of Engineering and Technology
Hindsthan University, TN	B.Tech IT (Cloud Computing) B.Tech CSE (Cloud Computing)	Faculty of Engineering and Technology
University of Technology and Management, Meghalaya	B.Tech CSE (Cloud Computing)	Faculty of Engineering and Technology

these institutes, crores of students are availing their education, skills, and training, etc. Surprisingly among these many are tagged as Engineering Institutes and such numbers are also good, i.e., 6375 according to the 2015 Data of AICTE. The All India Council for Technical Education (AICTE) is the governing body to control, grant, funding, and approved Engineering and Technical programs in India. But surprisingly not a single Engineering Colleges/ universality offer Cloud Comput-

Table 6 The Universities which are offering Cloud Computing in Masters level

Universities	Programs	Faculty/wing
VIT University, Tamilnadu	M.Tech-CSE (Cloud Computing)	Faculty of Engineering and Technology
Amity University, Uttarpradesh	M.Tech-Cloud Computing	Faculty of Engineering and Technology
Vel Tech University, TN	M.Tech (IT Infrastructure and Cloud Computing)	Faculty of Engineering and Technology
SRM University, Tamilnadu	M.Tech-Cloud Computing	Faculty of Engineering and Technology
KL University, Andhrapradesh	M.Tech-Cloud Computing	Faculty of Engineering and Technology

ing programs with their approval! Tables 5 and 6 depicted programs fall under the deemed/private universities for which AICTE not come under the jurisdiction.

Hence among the 6375 Institutes (with 1903722 intakes) not a single degree holder comes with Cloud Computing (*or related specializations like Virtualization, Big-Data Management and Cloud,* etc.) knowledge or skills. The Technical education also comprises (apart from Engineering) with the education of Architecture, Polytechnic, Management, Computer Applications, etc. [23]. And in all these segment huge intake capacity and institutes offering may be seen in Table 7. But not a single institute offers Cloud Computing either full-fledged degree or specialization. Though, potentialities exist with existing Departments of Architecture, Management, Computer Applications, Polytechnic under the AICTE to offer Cloud and Related Science, Technology and Management Programs.

Interestingly among the Science and Technology fields, the Computing is the apex one. Actually, at the Engineering Bachelor level, the Degrees are offered as B.Tech/BE-CSE and B.Tech/BE-IT. Hence compare to any other discipline in Engineering the Computing is higher. Moreover in the AICTE approved Institutes another program also fall under the Computing, i.e., MCA (Masters of Computer Applications). Degree wise it is Masters but equivalent to the B.Tech CSE/IT. Regarding the total number of institutes; 1469 are offers the program with the annual intake of 1 Lakh+. Hence huge output is in Computing and IT field. However if we consider the MSc programs which are not under the AICTE (*but deemed as an equivalent of B.Tech/BE-IT/CSE*) and offered in most of the Universities and PG Colleges then total each year intake will be 2.5+ Lakh. Surprising! This is highest among other branches available in India in the Science and Technology domain. But null in Cloud Computing Degree or Specialization; but with huge current and future market. However, many of the institutes are offering Cloud Computing as a Coursework paper in their Bachelors and Masters program such as at the IIT Ropar, IIT Delhi, IIT Kharagpur, Delhi University, Jadavpur University, IIEST Shibpur, IIT Indore, IIT Bhubaneswar, Anna University, Pune Universty, BIT Mesra, MIT, Manipal, etc. Though in each and every segment huge potentialities are there to offer Cloud Computing and related

Table 7 Total number of technical institutes and intake under AICTE and possible Cloud Computing inclusion

Streams	Number of institutes	Total intakes (as on 2015)	Remarks on Cloud Computing potentiality
Architecture	177	(Annual intake of 11070)	Special program on cloud architecture and cloud data center designing may be offered
Engineering	6375	(Annual intake of 1903722)	As depicted in Tables 8, 9, 10 and 11
Management	3217 (MBA)+600 (PGDM)	(Annual intake of 366439)	Techno-managerial programs may be offered
Computer applications	1469	(Annual intake of 110585)	Specialization may be offered in MCA/BCA
Polytechnic	4276	(Annual intake of 1308008)	Special thrust may be given in higher semester for Cloud Computing papers, etc.

areas with Bachelors, Masters, Doctoral level. The next subsection is depicted such information and possibilities.

3.1 Potential Programs

Cloud Computing may be offered with several nomenclatures and easily accommodate in the course catalog of the universities, colleges of many kinds. Cloud Computing in engineering segment may be offered as a full-fledged degree or as a specialization in the related engineering programs such as Computer Engineering, Computer Science and Engineering, Information Technology, Information Science and Engineering (*mainly offered in the State of Karnataka*), Communication Engineering, etc. It mentioned that Engineering Degree intake in India is over 19 Lakh, but Cloud Computing programs are nil. The possible Bachelor's Degrees in different subjects have mentioned in Table 8 along with Engineering college's potential and proposed programs.

Similarly Cloud Computing is also possible to offer as the specialization in the Engineering Colleges in the existing programs with the nomenclature of Cloud Computing (*or related such as Cloud Engineering, Cloud and Big Data Management, Cloud and E-Commerce Management, Cloud and Virtualization Systems,* etc.) and such possible nomenclatures have been listed in Table 9.

Table 8 The possible Bachelor's degrees in different subjects and institutions

Engineering/ Technology	Sciences	Commerce	Arts	Applied Arts and Sciences
B.Tech-Cloud Computing BE-Cloud Computing B.Tech/BE-Cloud and Virtualization Engineering B.Tech/BE-Cloud and Big Data Engineering	BS/BSc-Cloud Computing BS/BSc-Cloud and Virtualization Engineering BS/BSc-Cloud and Big Data Engineering	B.Com (Cloud Business and IoT Marketing) B.Com (Major) in Cloud and Internet Economics B.Com-Cloud and E-Commerce	BA (Cloud Economics) BA (IT Management with Cloud Economics) BA-Cloud and Digital Economics	B.Des (Cloud Architecture) B.Arch (Cloud and Virtualization Designing) B.Plan (Cloud Systems)

Table 9 The possible Cloud Computing specializations in Bachelor's degrees in several fields

Engineering/technology	Sciences	Arts
B.Tech CSE (Cloud Computing) BE-IT (Cloud Computing) B.Tech/BE-IT/CSE (Cloud and Virtualization Engineering) B.Tech/BE-IT/CSE/ECE (Cloud and Big Data Engineering)	BS/BSc-Computer Science (Cloud Computing) BS/BSc-IT (Cloud and Virtualization Engineering) BS/BSc-Computer Applications (Cloud and Big Data Engineering)	BA-Economics (Cloud Economics) BA-Management (IT Management with Cloud Economics) BA-Economics (Cloud and Digital Economics)

Naturally, the Cloud Computing also has the potentiality to offer in the Masters level in Engineering, Science, Commerce, Arts, Designing, Architecture, etc., as a full-fledged degree and as specializations in the related fields (See the Table 10). While, Table 11 depicted the possible and potential specializations in the related subjects of Computing and Information Technology.

One another program in Computing, which is also popular in India is MCA (Master of Computer Applications) which is available in a large number of institutes and huge intake (1469 Institutes with 110585 intake capacities according to 2015 data) may also grab the potentiality to offer the Cloud Computing and related programs. Some of the proposed nomenclatures have been depicted in Table 11. Similarly in Management Segment the most popular is MBA (Master in Business Administration) and available across 3217 institutes under the AICTE with the annual intake of 366439. Hence whether MCA or MBA the intake and available institutes are Huge! But the regarding Cloud Computing context, the total output is zero!

Hence there are healthy possibilities to offer Cloud Computing and allied programs in the Indian context as far as MCA and MBA. The proposed programs with nomenclatures have been depicted in Table 12. The programs may also offer for the Post Graduate Diploma Programs such as PGDCA or PGDBA or PGDM as Cloud Computing specializations.

Table 10 The possible Master's degrees in different subjects and institutions

Engineering/technology	Sciences	Commerce	Arts	Applied Arts and Sciences
M.Tech-Cloud Computing ME-Cloud Computing M.Tech/ME-Cloud and Virtualization Engineering M.Tech/ME-Cloud and Big Data Engineering	MS/MSc-Cloud Computing MS/MSc-Cloud and Virtualization Engineering MS/MSc-Cloud and Big Data Engineering	M.Com (Cloud Business and IoT Marketing) M.Com (Major) in Cloud and Internet Economics M.Com- Cloud and E-Commerce	MA (Cloud Economics) MA (IT Management with Cloud Economics) MA-Cloud and Digital Economics	M.Des (Cloud Architecture) M.Arch (Cloud and Virtualization Designing) M.Plan (Cloud Systems)

Table 11 The possible Cloud Computing specializations in Masters degrees in several fields

Engineering/technology	Sciences	Arts
M.Tech-CSE (Cloud Computing) ME-IT (Cloud Computing) M.Tech/BE-IT/CSE (Cloud and Virtualization Engineering) M.Tech/BE-CSE/IT/ECE (Cloud and Big Data Engineering)	MS/BSc-Computer Science (Cloud Computing) MS/MSc- IT (Cloud and Virtualization Engineering) MS/MSc- Computer Applications (Cloud and Big Data Engineering)	MA-Economics (Cloud Economics) MA-Management (IT Management with Cloud Economics) MA-Economics (Cloud and Digital Economics)

Table 12 The possible Cloud Computing specializations in MCA/MBA programs

Computer applications	Management sciences
MCA (Cloud Computing) MCA (Cloud and Virtualization Engineering) MCA (Cloud and Big Data Engineering) MCA (Cloud and Digital Economics) MCA (IT Management with Cloud Computing)	MBA (Cloud Management) MBA (Cloud Economics) MBA (Cloud and Digital Economics) MBA (Cloud and IT Management) MBA (Cloud and Information System Governance)

   Cloud Computing has huge potentialities and wonderful market around the world and also in India. Today most of the corporate are moving towards Cloud enable systems and Information Systems designing and development. Thus manpower development is very much important and realistic; thus preparation of manpower is very much important and urgent. The proposed programs no doubt will help better IT systems in India. The Cloud Computing specialized programs will also help for creating a faster Digital India and a solid Digital India.

3.2 Tenability of Hypothesis

- Cloud Computing is an important domain and deals with technology, management principles, and electronics—hence it is a pure interdisciplinary domain. *Thus Hypothesis 1 is fully substantiated*.
- Cloud Computing and wider research evolved new fields like more greener cloud and similar systems bring Green Cloud Computing, Cloud-enriched Information Systems (Green), Environmental Informatics, Energy Informatics, etc. *Thus Hypothesis 2 is wholly substantiated*.
- Cloud Computing is available with many International organizations and vendors rather than its availability as University-based Bachelors and Masters programs as far as India is concerned. *Thus Hypothesis 3 is validated*.
- There are many universities around the world offering Cloud Computing programs mainly in PG level rather UG programs with an internship, twining programs, etc. *Thus Hypothesis 4 is purely confirmed*.
- Cloud Computing in India mainly offered as a chapter or paper in the related programs in the Universities and IITs, etc. *And not only in IITs*. *Thus Hypothesis 5 is partially confirmed*.
- Cloud Computing has the wider potentiality to offer the program in Bachelors and Masters level in Science, Engineering, Technology, Commerce, Management, Designing, Architecture, Humanities, etc. and that have been depicted in this research. *Thus Hypothesis 6 is entirely substantiated*.

4 Findings

- Cloud Computing is an integral part of today's Information System and Management. Thus numerous services and applications have been found association with several cloud applications.
- Cloud Computing has quite a lot of gradients, tools and among these eco-friendly tools and gradients are treated important ones for promoting several service models and types and their solid integration.
- Cloud Computing is also known as Cloud Engineering, Cloud and Big Data Management, Cloud and E-Commerce Management, Cloud and Virtualization Systems, etc.
- Cloud applications may also social media-based and there are healthy potentially to integrate many Cloud apps in large information systems building and intelligent systems creation.
- Cloud Computing and Big data systems' become an integral part of most of the social and development events in almost all the countries, numerous applications and products and thus many universities are offering Cloud Computing and allied programs as a chapter or full paper in many universities in India.

- A very few institutes offer B.Tech and M.Tech in Cloud Computing or as a Specialization and all of them are categorized as Private or Deemed; though there are 6375, 1469, 3217 colleges in the field of Engineering, Computer Applications and Management Science respectively under AICTE.
- Not a single IIT, NIT, IIM, IIEST, IISER, NISER, etc.; the apex and most prestigious institutes offers Cloud Computing programs; but it is easy to start and operate.
- India holds 750+ Universities and around 40,000+ General Colleges (including HEIs) and many of them offer PG programs but not a single offers Cloud Computing either a full-fledged degree or as specialization.

4.1 Suggestions and Recommendations

- Cloud Computing programs need to be integrated into the Government offices and establishments in diverse sectors, health segment, educational settings, etc. Thus government needs proper planning, initiatives for cloud-based systems creation for making digital territory.
- Cloud Computing and allied tools, systems are rising rapidly and thus global IT market many ways depends on this system and computing platform. Hence Government, IT Association, Companies need to build a proper framework, issues, etc. for a real Cloud Informatics practicum.
- It is better to start programs on Cloud Computing and allied domain in the Universities, as they are autonomous in nature and may start the program as full-fledged degrees or as specializations as proposed in the tables of this research work.
- It is essential that AICTE Should take a proper step for starting Cloud Computing programs either full or partially as specialization in their Engineering Programs, i.e., B.Tech/M.Tech etc.
- The Management and Commerce, Humanities and Social Sciences Departments may also start Cloud Computing and allied programs such as Cloud and IT Management, Cloud and Digital Economy, etc., as proposed and provided in Tables 11, 12.
- For offering Cloud Computing specializations in the B.Tech/M.Tech-CSE/IT or other related subjects, the university may grab the potentiality with adjustment of Cloud-related papers, courses, etc., by the adjunct professor, guest professors, etc.

5 Conclusion

Cloud Computing is a valuable name in the Information Technology domain and also in Government, health, education, etc. due to its wonderful benefits. Worldwide the market for Cloud Computing is rising. Similarly, several new concepts had risen. Such as Cloud-Green Computing, Green Information Systems and most importantly Big

Data Management or simply Data Science applications. India is developing in many contexts worldwide and in IT India already captured huge success. Cloud Computing and Virtualization programs no doubt will help in better digitalization, and healthy Social and media applications. India holds 40,000+ HEI with numerous Government institutions but it is a fact that only a few are offered Cloud Computing. And all these are either private or deemed universities. India's IT and Computing *Bachelors (Engineering) (Engineering)* or equivalent with MCA/MSc the total intake touches 2.5 Lakh. So, huge Cloud Computing skilled manpower may be generated with proper designing of academic, professional, and research programs. The MPhil/PhD program may be started with healthy Foundation coursework on Cloud Computing and Research Methodology. And parallel Cloud Computing and allied programs may also be offered during the Research Work as audit courses and proper skill development to keep the knowledge and practical skills up-to-date. Hence as a whole individual and collaborative efforts are required from the players like UGC, AICTE, MHRD, etc., including steps from the respective Colleges and Universities. A true Digital India is in many ways possible to build by building healthy Cloud Computing systems and it is possible by offering Cloud Computing education and research programs.

Acknowledgements Special Acknowledgement to Prof. R. Buyya, Director, CLOUDS, University of Melbourne, Prof. Mrinal Kanti Ghose, Dean (Academics), Sikkim Manipal University for their information guidance etc.

References

1. Buyya, R., Ranjan, R., Calheiros, R.N.: Modeling and simulation of scalable cloud computing environments and the CloudSim toolkit: challenges and opportunities. In: International Conference on High Performance Computing & Simulation, 2009. HPCS'09, pp. 1–11. IEEE (2009)
2. Dikaiakos, M.D., Katsaros, D., Mehra, P., Pallis, G., Vakali, A.: Cloud computing: Distributed internet computing for IT and scientific research. IEEE Internet Comput **13**(5), 10–13 (2009)
3. Calheiros, R.N., Ranjan, R., Beloglazov, A., De Rose, C. A., Buyya, R.: CloudSim: a toolkit for modeling and simulation of cloud computing environments and evaluation of resource provisioning algorithms. Softw. Pract. Exp. **41**(1), 23–50 (2011)
4. Davenport, T. H., Prusak, L.: Information Ecology: Mastering The Information and Knowledge Environment. Oxford University Press (1997)
5. Gurbaxani, V., Whang, S.: The impact of information systems on organizations and markets. Commun. ACM **34**(1), 59–73 (1991)
6. Clemons, E.K.: Information systems for sustainable competitive advantage. Inf. Manag. **11**(3), 131–136 (1986)
7. Foronda, V.R.: Integrating information and communication technology into education: a study of the iSchools project in Camarines Sur, Philippines. J. Dev. Sustain. Agric. **6**(1), 101–113 (2011)
8. Harmon, R.R., Auseklis, N.: Sustainable IT services: assessing the impact of green computing practices. In: Portland International Conference on Management of Engineering & Technology, 2009. PICMET 2009, pp. 1707–1717. IEEE (2009)
9. Hooper, A.: Green computing. Commun. ACM **51**(10), 11–13 (2008)

10. Kettinger, W.J., Lee, C.C., Lee, S.: Global measures of information service quality: a cross-national study*. Decis. Sci. **26**(5), 569–588 (1995)
11. Paul, P.K.: Green information science: information science and its interaction with green computing and technology for eco friendly information infrastructure. Int. J. Inf. Dissem. Technol. **3**(4), 292 (2013)
12. Watson, R.T., Boudreau, M.C., Chen, A.J.: Information systems and environmentally sustainable development: energy informatics and new directions for the IS community. MIS Q., 23–38 (2010)
13. Karthikeyan, N., Sukanesh, R.: Cloud based emergency health care information service in India. J. Med. Syst. **36**(6), 4031–4036 (2012)
14. Kumar, K., Lu, Y.H.: Cloud computing for mobile users: can offloading computation save energy? Computer **4**, 51–56 (2010)
15. Paul, P.K., Chatterjee, D., Rajesh, R., Shivraj, K.S.: Cloud computing: overview, requirement and problem in the perspective of undeveloped and developing countries with special reference to its probable role in knowledge network of academic field. Int. J. Appl. Eng. Res. **9**(26), 8970–8974 (2014)
16. Melville, N., Kraemer, K., Gurbaxani, V.: Review: information technology and organizational performance: an integrative model of IT business value. MIS Q. **28**(2), 283–322 (2004)
17. Subashini, S., Kavitha, V.: A survey on security issues in service delivery models of cloud computing. J. Netw. Comput. Appl. **34**(1), 1–11 (2011)
18. Schmidt, N.H., Erek, K., Kolbe, L.M., Zarnekow, R.: Towards a procedural model for sustainable information systems management. In: 42nd Hawaii International Conference on System Sciences, 2009. HICSS'09, pp. 1–10. IEEE (2009)
19. Paul, P.K., Dangwal, K.L.: Cloud Computing Based Educational Systems and iits challenges and opportunities and issues. Turkish Online J. Distance Educ.-TOJDE **15**(1), 89–98 (2014)
20. Wang, D.: Meeting green computing challenges. In: Electronics Packaging Technology Conference, 2008. EPTC 2008. 10th, pp. 121–126. IEEE (2008)
21. http://www.siliconindia.com/magazine-articles-in/25_Most_Promising_Cloud_Computing_Companies-LUBO681173186.html. Accessed on 20 June 2016
22. https://www.ugc.ac.in. Accessed on 20 June 2016
23. http://www.aicte.org. Accessed on 20 June 2016
24. http://www.mhrd.gov.in. Accessed on 20 June 2016

Data Processing Framework Using Apache and Spark Technologies in Big Data

Archana Singh, Mamta Mittal and Namita Kapoor

Abstract The worldwide usage of Internet has been generating data exponentially. Internet has re-evolved business operations and its number of consumers. The data generation begins with the fact that there is vast information to capture and store. The rate of mounting of data on the Internet was one of the important factors in giving rise to the concept of big data. However, it is related to Internet but its existence is due to growing unstructured data which requires management. Organization stores this data in warehouses for future analysis. Besides storage, the organization also needs to clean, re-format and then use some data processing frameworks for data analysis and visualization. Hadoop MapReduce and Apache Spark are among various data processing and analysis frameworks. In this chapter, data processing frameworks Hadoop MapReduce and Apache Spark are used and the comparison between them is shown in terms of data processing parameters as memory, CPU, latency, and query performance.

Keywords Apache spark · Hadoop · Big data · MapReduction · Query
performance

1 Introduction

Data is a collection of facts that are persisted for reference and data analysis. It is an organization asset which is growing day by day at fast pace and it can be used for data analysis to produce valuable and useful information. The volume of data is growing at a phenomenal pace which is happening with maximum usage of Internet. This data is not formatted with fixed structure but it is varied which is adding a level

A. Singh · N. Kapoor
ASET, Amity University, Noida, Uttar Pradesh, India
e-mail: archana.elina@gmail.com

M. Mittal (✉)
G. B. Pant Government Engineering College, Okhla, New Delhi, India
e-mail: mittalmamta79@gmail.com

© Springer Nature Singapore Pte Ltd. 2019
M. Mittal et al. (eds.), *Big Data Processing Using Spark in Cloud*,
Studies in Big Data 43, https://doi.org/10.1007/978-981-13-0550-4_5

of complexity. Several challenges occur that includes, like persistent storage, share, search, analysis, and visualization. The drift to larger data sets is due to the added information that comes from analysis of a single large set of data as compared to separate smaller chunks with the same total amount of data, allowing correlations to determine and analysis on data like tagging, traffic condition, etc.

Some organizations after facing hundreds of gigabytes of data for the first time may trigger a need to reconsider data management options. For others, it may take tens or hundreds of terabytes before size becomes a significant consideration. Big data has high-velocity, volume and variety information organization assets that demand inexpensive, innovative forms of information processing for enhanced insight and decision-making. The volume of data is growing at a phenomenal pace which is happening with maximum usage of Internet. This data is not formatted with fixed structure but it is varied, adding a level of complexity. The World Wide Web and social media applications like Facebook, twitter, etc., are some of the major contributors in generating large sets of data. Large data is difficult to work with using traditional relational database management systems and desktop statistics, instead "massively parallel software running on thousands of servers." Big data can be relatively defined as the synonyms of large amounts of data which requires distributed or single data storage that enables parallel processing tasks and the system is relatively inexpensive. Big data definition varies depending on the potentiality of the organization managing the large set, and on the potentiality of the applications that are used to analyze the data set in its domain. The organizations store and generate data in digital form and maintain the data in detail for long. The collected data becomes an organization asset and data maintenance becomes organization duty. The intricate data analysis can improve organization decision-making and control risks, and precisely tailored applications or services. We found that the decision-making process consumes maximum effort and time in querying and responding of data. Large efforts are required in data transfer from one location to another. The massive efforts in querying and transfer resulted in inefficiency of the system. Non-operational persisted big data can later be used in designing and developing next generation applications and tools. Telecom organizations are using data obtained from VAS services to leverage innovative ideas and services as per customer needs.

In this chapter, the data processing techniques in big data are discussed and used. A Hadoop ecosystem offers different tools to process the data. These tools are broadly categorized as MapReduce sub systems and NonMapReduce sub systems. Hive, Pig, Mahout, Sqoop, and Flume, etc., tools comes under the category of map reduce sub systems whereas, Spark, Impala and HBase, etc., comes under non-map reduce sub systems. The chapter discussed the architecture design of Hadoop and Apache Spark and compared data process/visualization approach using map reduce and non-map reduce tools. The key parameters for comparison is used are query performance, latency, CPU, and RAM evaluation of different sizes of data.

The chapter is organized as follows: In the following section related work done is explored. Big data as a revolution is discussed in Sect. 3, and Hadoop and Apache Spark architecture are explained. In Sect. 4 the experimental comparative analysis between Apache Spark and Hadoop and results are presented. In Sect. 5 conclusion is mentioned.

2 Related Work Done

Dr. Urmila R. Pol explained the set of execution commands in MapReduce, Pig, and Hive. The author has executed word count program on sample input data on each frameworks. Author studied the comparison among MapReduce, Pig, and Hive [1]. Lei Gu, Huan Li conducted exhaustive experiments to evaluate the system performance between Hadoop and Spark using typical iterative algorithm PageRank. Experiments were performed on cluster architecture. The experimental cluster is composed of eight computers. One of them is designated as master, and the other seven as slaves. Author studied running time comparison between Hadoop and Spark using real and synthetic graph datasets [2]. Shi, Xuanhua, Chen, Ming, He, Ligang, Xie, Xu, Lu, Lu, Jin, Hai, Chen, Yong and Wu observed the memory management and static memory size problem for Map or Reduce task [3]. Amol Bansod exploited the Apache Spark framework for efficient analysis of big data in HDFS, and compared it with other data analysis framework—Hadoop MapReduce. The author has represented comparison study for two iterative algorithms K-means clustering logistic regression between MapReduce and Spark [4]. Jai Prakash Verma and Atul Patel have conducted experimental analysis on customer review and feedback dataset. Author run word count program with different sizes file on Spark and MapReduce and recorded execution time in each framework [5].

3 Big Data as a Revolution

The world is an experiencing a massive data revolution or "data deluge". Whereas few years back, a relatively different media, small volume of data in comparison of today's data, was processed and made available on different channels. There are various sources on different channels through which such massive amount of data is generated in today's world. It is projected to increase by 40% annually in the next few years which is about 40-folds. The below Fig. 1, graph states increasing data rate of previous years as well as expected increase of coming years which is shows an increase of more than 40-folds between 2008 and 2020. This led to the rise of big data.

Internet has re-evolve business operations and its consumers. Then comes big data, which starts with the fact that there is vast information to capture and store. Growing data on Internet was media for introducing big data concept. However, it is related to Internet but its existence is due to growing unstructured data which would require manageability. Social media was one of media for rise in data. Study shows worldwide figures of leading social networks till July, 2013 (Figs. 2, 3, 4, 5, 6, 7, 8, 9, 10, 11 and 12).

The study shows number of social networks users in India from 2011 to 2012 with projections until 2017 (Tables 1, 2 and 3).

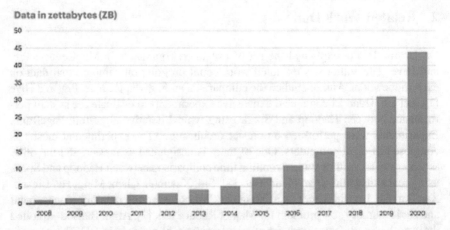

Fig. 1 Data rise till the year 2020. *Source* Internet

Fig. 2 Data generated on social media till 2013

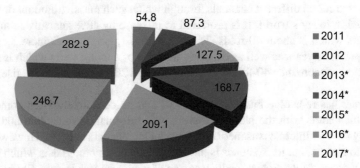

Fig. 3 Data generated on social media till 2017

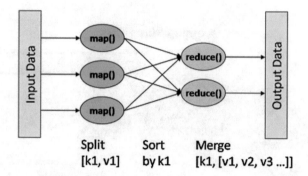

Fig. 4 MapReduce

HDFS Architecture

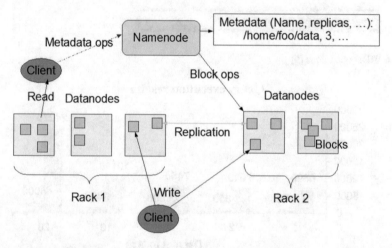

Fig. 5 Hadoop architecture [6]

Data can be stored in data warehouse and can be referred later for reporting and visualization. Data warehouse is also known as a database which stores heterogeneous non-operational organizational data. Here, the term heterogeneous means disparate data and the term non-operational means no update on data, i.e., service and audit logs, etc. It also persist historical data. One can use this data in further analysis to help in strategic decision in business. This data may contain errors and noise and it can be inconsistent which can make data warehouse integration complex. Data needs to pass through ETL (extract transform and load) cycle before it can be saved into data warehouse. This cleaned and transformed data can be stored at traditional data warehouse or hadoop, etc., herein, we are considering that data is stored in Hadoop eco system.

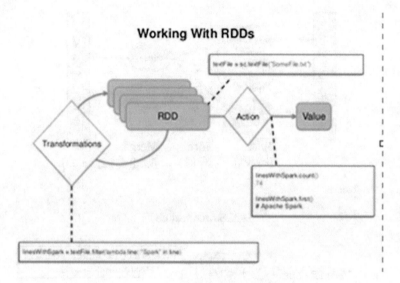

Fig. 6 RDD operations [6]

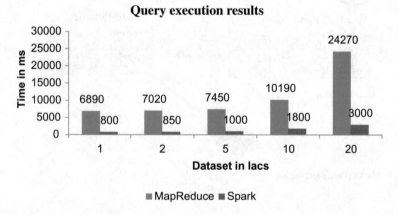

Fig. 7 Query execution results

The technology has V's to define its nature. Three V's associated with big data
Volume—Volume defines the amount of data generated. High volume of data is big
data. Every organization will need to find a way to handle growing volume of data that
is increasing every day. We certainly find this happen within our organization—data
volume is getting increased day by day.
Velocity—Velocity defines the frequency at which big data is generated, gathered,
and shared. Recent developments shows that not only end customers but also business generate more and more data in shorter time. Seeing the fast speed of data,

```
hadoop fs -put /tmp/household_power_consumption.csv /user/maria_dev/data/

CREATE EXTERNAL TABLE IF NOT EXISTS powerdataset(powerdate  VARCHAR, powertime

VARCHAR, active_power DECIMAL, reactive_power DECIMAL, voltage: DECIMAL,

intensity DECIMAL,metering_1 DECIMAL,metering_2 DECIMAL,metering_3 DECIMAL)

ROW FORMAT DELIMITED FIELDS TERMINATED BY ',' STORED AS TEXTFILE location

'/user/maria_dev/data/household_consumption.csv';

Set mapreduce.input.fileinputformat.split.minsize=10485760; --10MB

Set mapreduce.input.fileinputformat.split.maxsize= 20971520; --20MB

select      powerdate,     sum(active_power*1000/60-metering_1-metering_2-

metering_3) energyperminute from powerdataset group by powerdate;
```

Fig. 8 Hive query

```
hadoop fs -put /tmp/household_power_consumption.csv /user/maria_dev/data/

val dataset = sc.textFile("/user/maria_dev/data/household_power_consumption.csv")

case class datasetclass (powerdate:     String, powertime: String, active_power:Double,
reactive_power: Double, voltage: Double, intensity: Double,metering_1:Double,metering_2 :
Double,metering_3: Double)

valcolmapping=dataset.map(k=>k.split(",")).map(k=>datasetclass(k(0),k(1),k(2).toDouble,k(3).t
oDouble, k(4).toDouble, k(5).toDouble, k(6).toDouble, k(7).toDouble, k(8).toDouble))

colmapping.toDF().registerTempTable("powerdataset")

select      powerdate,     sum(active_power*1000/60-metering_1-metering_2-metering_3)
energyperminute from powerdataset group by powerdate;
```

Fig. 9 Spark SQL

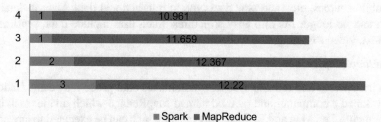

Measured latency (in seconds)

	Spark	MapReduce
4	1	10.961
3	1	11.659
2	2	12.367
1	3	12.22

Fig. 10 Latency in Spark and Hadoop

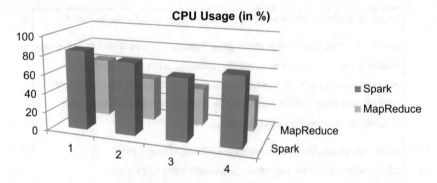

Fig. 11 CPU usage in Spark and Hadoop

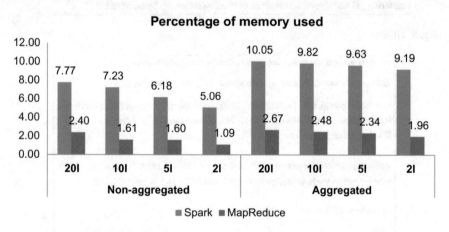

Fig. 12 Percentage of memory used by Spark and Hadoop

organizations can only capitalize on this data if it is gathered and shared in real time. Now there are many analytics and CRM or similar systems fall.

Variety—A rapid growth of data from many sources, i.e., social, web application, and mobile sources, etc., add new data type to transactional data. Data is detailed enough and no longer fits into easy structures. Such data include data, sentiment, social text, video steaming, and web, etc.

A. *Hadoop MapReduce*

This is transparent and reliable system which makes data processing easy. Hadoop has developed a computational method named MapReduce which divides task into two fragments, i.e., Map and Reduce. Any one of which can be executed to any node as instructed by master scheduler. In addition, it also provides a distributed file system that stores data on different multiple nodes. MapReduce and distributed filesystem are designed to handle nodes fault tolerance automatically. It enables application to work on large data. Hadoop uses MapReduce engine to compute statement by distributing task across nodes in a cluster. MapReduce algorithm states that for any ongoing task Map and Reduce are independent to each other and runs parallel to each other on different keys and data. Map performs data selection from individual node and reduce combine and data sets to sends it user. We can run Map operation on machines running nodes where data lives. Each node pushes data to the distributed filesystem in machines. The final output can be achieved when reducers run and combine the results. There can be multiple reducers on each machine in case of distributed cluster networks of database. Each enter may reduce different keys as well. Map reduce working (source: DZone Community).

The Map

Master node accepts user's input. Name node at master node decomposed user's input into smaller chunks. Name node then distributes the tasks among worker nodes. Node manager at each worker node processes the request. After completing execution task tracker returns the output to master node which resides at master node. A map function provides output in key-value pair where key is the key-value pair of input row and value is the key-value pair of resulting rows.

Map (key1, value) ->list<key2, value2>

Output value can contain multiple entries in case multiple matching records are found.

Shuffle and Sort

MapReduce ensures sorting and shuffle, i.e., it ensures that output of map function when passes as an input to reduce method is key-wise sorted. This process of

Table 1 User-generated data on social media till 2013

Leading social networks	Number of registered users in millions as of July 2013
Facebook	1155
Twitter	500
GooglePlus	500
LinkedIn	225

Table 2 Number of users using social media till year 2017. *Source* www.statista.com

Year	Number of users in millions
2011	54.8
2012	87.3
2013	127.5
2014	168.7
2015	209.1
2016	246.7
2017	282.9

Table 3 Computation time
of query processing

No. of records in lacks	MapReduce in milliseconds	Spark in milliseconds
1	6890	800
2	7020	850
5	7450	1000
10	10190	1800
20	24270	3000

sorting map outputs and transferring to reducer is known as shuffle. Shuffling is an intermediate process which occurs after Map and just before Reduce function.

The Reduce

Map and Reduce functions run in parallel and are independent to each other. This parallel processing is very effective for large data. We can compare MapReduce together with SQL query Select with Group By. A reduce function accepts output of Map function as input and generate aggregated output again in key-value pair.

Reduce (key2, list<value2>) -> list<value3>

The MapReduce Engine

MapReduce algorithm states that for any ongoing task Map and Reduce are independent to each other and runs parallel to each other on different keys and data. Map performs data selection from individual node and reduce combine and data sets to sends it user. We can run Map operation on machines running nodes where data lives. Each node pushes data to the distributed filesystem in machines. The final output can be achieved when reducers run and combine the results. There can be multiple reducers on each machine in case of distributed cluster networks of database. Each enter may reduce different keys as well. Distributed file system maintains multiple copies of data across different machines. Spreading data at multiple machines offers reliability with having maintained RAID controlled disk at machine. It instructs multiple nodes to perform Map operation. If any of the nodes is found to be non-working, it requests another backup node to fulfill the operation. Yarn Scheduler keep an eye on MapReduce jobs execution, map operations, and respective reduce operations. Yarn resource manager also monitors success and failure of these tasks and executes entire batch. These filesystem and yarn resource manager is accessible by the programs that are allowed to read and write data as well as data submission after MapReduce jobs. Apache has invented MapReduce engine. Its aim was to provide distributed file system for large database with failover mechanism as well.

The request of data is when a client request for certain data, this request goes to Namenode. Nodemanager then tracks map operation and combine operation as well as input and output path of data at datanodes. The yarn application manager will determine number of splits and datanodes on different machines. Yarn scheduler then send task to different data nodes. Application master starts executing Map operation at datanodes and extracts data. When map finishes task, it notifies to Node manager.

When all the map tasks are done, they notify yarn at master node. Yarn resource manager then chooses datanodes for reduce function and notify all datanodes for reduce function. Each container at datanode starts reading remote files and sorts data on key-value pairs. It starts reduce job and starts aggregating data on key.

This framework is designed such a way to make it crash free. The application master tracks of each phase progress and keeps on pinging datanode to check its status. As soon as any datanode crashed, the namenode will reassign the map task to different datanode. As map and reduce methods completes, the namenode resumes client program.

Advantages

Map Reduce advantages over conventional database are given below.

1. It is easy to use and implement. Programmers can work without having any prior knowledge of distributed systems. It encapsulates details like fault tolerance, parallelization, etc., from programmers. This is language free so programmers are free to use language of their choice, i.e., Java, php, etc.
2. It can handle large variety of computation problems easily. Map Reduce is used in data mining, searching and sorting, etc.
3. It enables applications scaling across large clusters of machines connecting multiple nodes having fault tolerance built in feature for better performance.
4. It offers automatic parallelization with the help of different trackers and nodes.
5. It handles fault tolerance.
6. It enables inter machine communication.

Limitations

1. We can minimize computing time by running Maps and Reduces parallel. We cannot achieve desired throughput with serial execution of MapReduce. We cannot even control the order of map and reduce methods.
2. It becomes inefficient when repeated search operations are being performed again and again.
3. Map and Reduce runs in sequence, i.e., Reduce method will not start until all maps operations are completed. This sometimes results long waiting time.
4. It is assumed that reduce results will be smaller that map results which means It fetches large output and generating smaller ones to end user.

B. Apache Spark

It is a framework for data processing. It uses Hadoop as storage engine and manages its own cluster management for data computation. Unlike hadoop map reduce it computes data in memory. As known that map reduce stores each computation output into distributed files but in contrast to this feature, spark does not store each computation output in any file system but it keeps all processing output in memory.

Unlike Hadoop it works well with iterative algorithms, batch processing, etc. It provides core API in many programming language. Writing a code in programming languages like Java and Scala, etc., is quite easy than writing a code for map reduce.

Spark also supports other libraries like SQL, machine learning library (mlib), and library for graph computation (GraphX), etc. It has two main features.

RDD (Resilient Distributed Dataset)

It is the key feature of Spark. It is collection of objects stored into different partitions. These objects are immutable thus whenever these are modified, a new RDD is created leaving original RDD in same state. It is fault tolerance because it knows when to recreate and when to recompute the dataset. It is the core of Spark which helps with rearranging the computation and optimizing the data processing. RDD working (source: databricks).

RDD Operations

Transformation:

This operation neither evaluates any dataset nor returns any value. Return type of transformation functions are RDD only. Transformation functions are groupbykey, map, pipe, flatmap, and reducebykey, etc.

Action

This operation process RDD and evaluates data value. When these functions are called, they compute data processing queries and return dataset values. Actions functions are count, first, reduce, countbykey, collect, count, first, take, and foreach, etc.

Directed Acyclic Graph

As name suggests it represents sequence of nodes and edges where node is an representation of each RDD partition and edge represents data transformation between nodes. It reduces multistage execution in hadoop model therefore enhances performance over hadoop.

Spark Architecture

It has master slave architecture with two daemons, i.e., master daemon and worker daemon also named as driver process and slave process respectively. It too has cluster manager which manages spark cluster. There is a single master daemon and any number of worker daemons. The key points can be discussed as follows.

Master daemon is the entry point of Spark Shell. It is responsible to create Spark context. It contains components like DAGScheduler, BackendScheduler, TaskScheduler, and BlockManager. It translates user instructions set into spark jobs and schedules its execution. For scheduling job execution, it requests required resources from cluster manager. It also converts RDDs into DAG and schedules. Worker distributes task to executors. Number of executors are configurable. User can predefined executor instances. Executor process the given task and returns output. Output can be in-memory or hard disk storage. Cluster manager is responsible for acquiring required resources for job execution. It allocates resources like memory, etc. It manages different scheduled jobs.

Advantages

Its in-memory cluster computing executes fast batch processing.

Besides core API, It also offers more libraries like GraphX, Spark SQL, spark Mlib (machine learning library), etc.

1. It is good for iterative and live stream processing.
2. It is fault-tolerant and flexible.
3. It can run on Hadoop ecosystem just like Hive and pig etc.

Limitations

Because of in-memory computation, it consumes large memory. Sometimes, cause memory-related issues.

1. It does not have its own file management system thus uses Hadoop HDFS or some other system.
2. It uses extensive system resources.

4   Experiment and Results

We have performed comparative study on Apache Spark and Hadoop frameworks with the help of dataset. The analysis was done on household electric power consumption dataset of approx. 2 million records. Dataset attributes are powerdate, powertime, power_active_power, reactive_power, voltage, intensity, metering_1, metering_2, metering_3

Sample records are

26/6/2008,23:26:00,0.306,0.212,242.77,1.4,0,1,0
16/12/2006,18:20:00,2.928,0,235.25,12.4,0,1,17

System configuration

OS: centos6 (x86_64)
Cores (CPU):2 (2)
Memory: 8.0 GB

Query Performance

We performed several queries on different size of datasets. We chose hive as MapReduce ecosystem tool and Spark sql as non-reduce ecosystem tool. MapReduce reducer size was set to 20 MB in hive environment. We studied that MapReduce takes around 24 s in processing 2 million records. For processing 2 million records, MapReduce execution is shown in below figure.

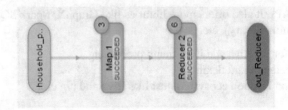

Stats	
Succeeded Vertices	2 Succeeded
Total Vertices	2
Succeeded Tasks	9 Succeeded
Total Tasks	9
Failed Tasks	0
Killed Tasks	0
Failed Task Attempts	0
Killed Task Attempts	0

On other side, we have found that spark takes only 3 s in processing 2 million records. Sample data outcome graph is shown below.

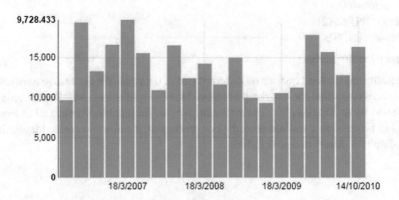

5 Comparison Between MapReduce and Apache Spark Framework

Above data stated that data processing in spark is fast in comparison of hadoop mapreduce task [7]. Graphical view is shown in below figure

Hive Query

See Fig. 7.

Spark SQL

See Fig. 8.

Latency

We monitored latency in spark and Hadoop MapReduce, respectively. We performed data computation on raw HDFS data and save the output back to HDFS at different location. We used data frames and RDDs in spark whereas, we have written MapReduce program in java. We run different cycles of data computation snippets with different data sizes using Spark and MapReduce respectively. Spark snippets used two executors and MapReduce run on single node. Latency was measured each time with each mode. We found that spark performs data computation with low latency. Besides recording latency, we have also observed CPU consumption of above executions. CPU usage is shown in figure given below.

RAM Evaluation with Aggregated and Non-Aggregated computations

Memory usage is one of the main checkpoints in data processing and analysis. Aggregation requires shuffle and reduce. Spark says that lower the number of partitions higher the shuffle block size. Block size beyond 2 GB is not recommended. Recommended partition size is 128 MB. We run our iterations with default partitions, i.e., 200 and found that spark consuming high RAM than MapReduce. One of the main reasons is execution memory. Spark keeps shuffle data in memory until group by or reduce by operation executed. It also needs memory to store the cache data to the disk. On the other side, MapReduce uses less memory because it writes intermediate (map job) results into disk first and thereafter, reduction cycle is performed. We have recorded the outputs of different cycles which are shown below.

6 Conclusion and Future Scope

Big data is offering new directions for data analytics collected from social frontier. The amount of information and knowledge that can be extracted from the digital media is can be leveraged to expand business. Such data is very informative and can be helpful in finding new ways to manage and process data. Undoubtedly, big data was found as a challenge in the beginning but visualizing its use cases has made it era of research and innovations. In this chapter, we studied Spark and MapReduce and presented brief comparative analysis between Apache Spark and Hadoop MapReduce frameworks [8]. Comparison analysis was done on dataset and this exercise was done to measure some of the key differentiators between Apache spark and Hadoop MapReduce. Besides finding differences, we have studied the similar features between both frameworks like scalability; fault tolerance; high level abstraction and both can process HDFS data, etc. We have also studied basic significant differences between Spark and MapReduce like code writing is lengthy in MapReduce. Spark can process interactive data which is not supported in MapReduce. The chapter presented comparative analysis on key features as query performance, latency, CPU, and RAM evaluation on different sizes.

This work can be extended further and we can study and compare more features like machine learning library, etc. Solution of existing big data issues and challenges of big data world will require extensive research work. Data processing on large data set of discrete data is found challenging and processing in terms of finding high valued data out big data requires efforts and study.

References

1. Gu, L., Li, H.: Memory or time: performance evaluation for iterative operation on hadoop and spark. In: IEEE, 978-0-7695-5088-6, Nov 2013
2. Shi, X., Chen, M., He, L., Xie, X., Lu, L., Jin, H., Chen, Y., Wu, S.: Mammoth: gearing hadoop towards memory-intensive MapReduce applications. IEEE 26(8) (2015). ISSN 1045-9219
3. Bansod, A.: Efficient big data analysis with apache spark in HDFS. Int. J. Eng. Adv. Technol. (IJEAT) 4(6) (2015). ISSN: 2249-8958
4. Verma, J.P., Patel, A.: Comparison of MapReduce and spark programming frameworks for big data analytics on HDFS. IJCSC 7(2) (2016)
5. Shi, J., Qiu, Y., Minhas, U.F., Jiao, L., Wang, C., Berthold: Clash of the titans: MapReduce vs. spark for large scale data analytics. VLDB 8(13) (2015)
6. Hadoop MapReduce documentation. http://hadoop.apache.org/
7. Spark documentation. https://spark.apache.org/documentation.html
8. Pol, U.R.: Big data analysis: comparison of hadoop MapReduce and apache spark. IJESC 6(6) (2016). ISSN 2321 3361 © 2016

Implementing Big Data Analytics Through Network Analysis Software Applications in Strategizing Higher Learning Institutions

Meenu Chopra and Cosmena Mahapatra

Abstract With the growth and large-scale usage of social networks in the dissemination of knowledge in Higher education Institutions, a need is being increasingly felt to tame and utilize this vast (read Big) data for a more worldly use. With the use of various NASA (Network Analysis Software Applications) Tools, this aim can be easily achieved. NASA can be applied to various online social media networks generated data sets used by Educational Institutions like Twitter, Linked, or Proprietary Institution specific platforms for predicting and formulating student specific academic, safe-campus, and business strategies. As widely known, the above-listed social media (SM) applications help us in sharing digital artifacts (antiques) and capturing digital footprints. The common utility among these social media applications is that they are able to create natural network data. These OSMNs (online social media networks) represent the links or relationships between content generators as they look, react, comment, or link to one another's content. There are many forms of computer-mediated social interaction which includes SMS messages, emails, discussion groups, blogs, wikis, videos, and photo sharing systems, chat rooms, and "social network services". All these applications generate social media datasets of social friendships. Thus OSMNs have academic and pragmatic value and can be leveraged to illustrate the crucial contributors and the content. Our study takes all the above into account and thus shall explore the various Network Analysis Software Applications to study the practical aspects of big data analytics that can be used to better strategies Higher learning Institutions.

Keywords Online social media networks (OSMNs) · Network analysis software applications · Gephi · Graph · NodeXL · Pajek · SNM

M. Chopra (✉) · C. Mahapatra
VIPS, GGSIPU, New Delhi, India
e-mail: meenu.mehta.20@gmail.com

C. Mahapatra
e-mail: cosmenamahapatra1@gmail.com

© Springer Nature Singapore Pte Ltd. 2019
M. Mittal et al. (eds.), *Big Data Processing Using Spark in Cloud*,
Studies in Big Data 43, https://doi.org/10.1007/978-981-13-0550-4_6

1 Introduction

Online Social Networks (OSN) can be formed in any of the social networking sites like Twitter or Facebook[1] and can shaped by anything like classrooms, family, project, or sports teams. The nodes are made by real-world entities and the relationships between these entities are called the edges between nodes. The graph network becomes more complex by adding more nodes and edges between them.

Online social networking is something that the vast majority of individuals think about it, however, they do not think about the analysis of OSN, which is important in many domains like academia. Social Network Analysis—SNA is an Endeavour to answer a few questions, for example, which members or users are pioneers and which of them are followers? Who all are the powerful members? Are there any communities or groups and how they are framed? Which members are essential in a group? Which users are the exceptions or outliers? Which connections are vital? Initially, OSN is small, so for that analysis can be simply done manually, which is a simple process, but impossible with vast and complex systems, so when the system is substantially big like the real-world social networks for that Network Analysis Software Application (NASA) can be used.

The OSN are characterized as a network of connections or relationships or collaborations, where the nodes of the network comprise of individuals and the edges comprise of the connections or collaborations between these online users or actors [1]. The OSN and its tools or techniques to explore them existed for decades [2]. There can be a few types of OSN like coloration, email, or phone networks, but currently, the online social networks like Twitter,[2] LinkedIn, Facebook and so forth have been developed which gained the popularity by having millions users within short span of time. The field of Social Network Analysis (SNA) has evolved from statistics, graph theory, and measurements and it is used in few different application areas like data science, education, economy, business applications, etc. As the topology of the graph is similar to the real-world social network, so therefore, to analyze a social graph with NASA or graph analysis tools can be achieved. The drawback with graph analysis tools is that they cannot handle today's such real-world big and complex data sets or networks which contains a large number of hubs and edges. Moreover, they are dynamic in nature, i.e., there is no constant size, and nodes have multiple attributes and small- and large-sized communities formed within these networks.

In this chapter, some tools are examined, explored, and compared for online social network graphs. We have considered a few tools in particular Gephi1,[3] Graph, NodeXL,[4] and Pajek.[5] We have included more similar tools with their outcomes focusing on graph properties, effectiveness, efficiency, and visualizations highlights.

---

[1] www.facebook.com: Very famous social networking site.

[2] www.twitter.com: Social networking site.

[3] https://gephi.org: Social network analysis tool.

[4] http://www.smrfoundation.org/nodexl/: Microsoft Excel add-in for social network analysis.

[5] Pajek vlado.fmf.uni-lj.si/pub/networks/pajek: Social network analysis tool.

2 Using SNA and SNM in SNS

Social Networking Sites (SNS) are defined as a web-based application that introduces users to create a public or controlled profile in a large and systematic system as well as sharing similar interest within a shared network [3]. The beginning of their infant years on the basis of documents sharing, to finding old classmates online, to its ultimate boom through applications such as Twitter, Facebook, Myspace[6] and many others, SNS have been part of the current generation's agenda [4].

SNS require the user to sign up for services in which certain important and basic information needs to be given to the system. This is the basic way of segmenting the user in the system. Users then creating a "public" profile that can be viewed by others in the similar network. Privacy issues have been rather overlooked due to the euphoria of the ease of connection and interaction these SNS support. Dwyer et al. has perform a comparison between Facebook and Myspace, found that in online interactions, trust is somewhat less important in the creation of relationships as compared to in the offline world [5]. Once a user starts joining a network, the user starts to create extended relationships. The more friends they have, the more updates they will get [6]. The basic rule of SNS is that the user's level of activity within the network has an effect on the size of the user's personal network.

Social network activities are however never static. The movement of the network has created changes in SNS [7]. For an organization such as a university, this factor can lead to a great discovery that can be manipulated especially finding hidden behaviors of students.

2.1 The Attractions of Social Network Sites

There are several factors that attract students to communicate and sharing knowledge online through SNS such as quick notification responses and self-expression opportunity. Therefore, forum and wall-to-wall discussion on SNS are two popular features that could be the most effective ways for online group studies. These features drive toward knowledge sharing culture among the students. They are motivated to share their photos, videos, notes, and even their unprompted thoughts and opinions. It is suggested that SNS could be one of effective tools to cultivate a creative group work and constructive feedback in the higher learning environment.

2.2 Social Network Analysis: (SNA)

SNA is the methodical analysis of the OSNs. By using SNA, we can examine social connections in terms of network dynamics, comprising of hubs (representing nodes

---

[6]www.myspace.com: Social network interactive site.

within the network) also, ties (which represents the connections between the people, for example, follow–follower, family relationship, authoritative position, formal connections, and so on). Few things which we discover after doing an analysis of OSN are

1. Finding the structure of the OSN.
2. Finding different network statistical values for the social graph For example, graph diameter, CC (Clustering Coefficient), Centrality, transitivity, Strong and Weak Communities within the network, average degree, number of nodes and edges, betweenness, modularity, shortest path and so forth.
3. Detection of large and small communities in the OSN.
4. Graphical representation of the entire or part of the social network.

A few works have been done on different OSN to analyze and study different kinds of individual behavior patterns and their respective relationships with other members of the network [8–11].

There are two kinds of SNA, first, complete network analysis (CAN)—is analysis of connections among an arrangement of nodes in a social network. For CAN, the methods used are comparability analysis (equivalence, subgroup analysis and network measures like centrality (betweenness, closeness and degree) all require full complete network [12]. Second, ego or personal network analysis (ENA)—is the analysis of individual actors' behavior and its variation described and mined. A social network can have the many numbers of egos as nodes, which can be people, associations or entirety society in the social graph. The process of finding hidden elements or properties in a social network is called as Social Network Analysis (SNA). Many researchers conduct the analysis with the interest from various aspects. According to Brandes, social network consists of relationship between nodes or actors in a network where it is theoretically modeled as a graph, $G = (V, E)$, V is the actors and E as relations or potential attributes.

Through SNA, a particular organization can have a deeper understanding of the activities or events that are that happen around its community. For example, in a college circle, the network can be analyzed through the relationship between the students to students and between the students to their lecturers. However, other external factors may involve like getting interactions from other staffs of various departments. This leads to a more complex and extended version of the relationship if details of the social network are derived continuously.

2.3 Social Network Mining: (SNM)

Social Network Mining (SNM) is the implementation of data mining on social network sites. Information posted on profiles is mined to identify hidden data that can be manipulated. For example, mining keywords posted by users on profiles or status can lead researchers to identify the users' activities and whereabouts.

SNM has been used to find the hidden patterns or important information over large data size. This includes database consisting associated objects like people, events, organizations, and other possible objects that may relate to each other [13]. SNM is also applicable to other organizations. For example, in government sectors, SNM can be used in intelligence activities. This can be considered as a part of defensive approaches where a country can create an expectation of what kind of threats that might strike the country. For example, in the United States, the concern over terrorist attacks has increased the usage of network mining in its defense system [14].

3 Dataset Used

Our research paper titled "Education Gets the Facelift by Going Social" [15] shows the existence of crucial folks in the social media networks and also the relationships that exist between various social networks and their posts. To better understand the structure of online social media networks, we used NETVIZZ[7] application [16] to collect the dataset of the online social networking group from the popular social networking site, i.e., Facebook: named -Vivekananda School of Information Technology, it is the "closed" public group, which shared posts, comments, and members are visible to every other Facebook user. It has present 367 members and 6468 edges, are dedicated to sharing information about education, faculties, higher education professionals or related to academics.

4 Network Analysis Software Applications (NASA)

SNA can be achieved by using NASA—which are used to analyze, visualize, identify, or recognize nodes (learning, knowledge, organizations) and edges (association or relationship or connections) from different sources and format of data including scientific or mathematical models of OSN. There are a many applications available for accessing and analyzing these social graphs. Our aim is to exploit various NASA tools for "Big Data" manipulation, visualization and to show the social media network analysis of the most of the population of scholars and other users to make analysis easier. While many other traditional or empirical methods use programming languages for network analysis simply need domain knowledge, technical skills, and on-hand experience to explore the online media datasets. As network science moves to less computational and algorithmic-related areas an urgent need or demand for non-programmatic or graphical event-based interfaces are required. These various NASA are used for sharing or fetching the dataset which can be used further for analysis purposes to understand the resulting dataset. Now in the current scenario, the focus is shifting to the network analysis of OSNs repositories starting from public

---

[7]NETVIZZ: Tool used to analyze Facebook data.

open-ended discussions, common media sharing systems to the personal networks. Our study tried to explore few network analysis applications to analyze social media network data sets. Nowadays, many Internet social network services are available to store personal communications like email, instant message and logs comprise chatting data and shared files. The analysis that is done for the media likes Facebook populations and artifacts can portray a picture of the aggregate structure and perspective of a user's social world.

NASA can answer a few questions like what predictions or patterns can be generated by the user's interactions in a social media network? How are the users connected to other members in the group? Who plays critical roles like the commentator, friend, post author, or ego? What are the possible user demographics, chat conversations, or discussion topics or phrases that have captivated the most interest from different users on a network? How do communities behave within the social media network? Our goal is to explore NASA, which is simple, easy to use (i.e., Does not require any employment of technical programming language (source code)) for data analysis and visualization and can be adopted by a large community of researchers to experiment OSN. The NASA has rich graphical interfaces (GUI), vast libraries of metrics and it does not need coding or command line execution features [17, 18].

There are two kinds of NASA, one package for GUI and second, packages for programming language or scripting. Those NASA, which are based on GUI format are easy to learn, extensible, and powerful, well documented and mostly used as Gephi, Node-XL, Pajek, so on, whereas programming applications or packages are good in performance, well documented but difficult to learn. Various NASA are used to calculate network statistics and properties (for example, radii, diameter, CC, Betweenness, shortest path, community detection) by creating edges and node list, adjacency matrix, graph visualization, etc. By graph representation or visualization, it is easy to understand network dynamics and it also facilitate to detect timelines, relationships, strong and weak communities, isolated nodes, etc.

Data can be accessed from OSNs for analysis in the following ways; (i) Direct database exposure [19], (ii) Using API (application programming interface) for example, Graph API application on Facebook, (iii) Using Crawling agents for e.g. spiders or bots [20], all these mentioned techniques requires programming or manual interaction. Today many network analysis software applications tools (third-party tool that integrates with platform) are available with proper sanctioned APIs for data extraction and gathering of data some examples are (i) NameGenWebi which exports dataset that contains several variables to support user's, their friends and friendship network, (ii) a plug-in for NASA-NodeXL is "Social Network Importer", which can use for the analysis and visualization toolkit created by the group of scholars which provides features like extracting extensive data (it includes posts, comments, and other activities) for all kinds of graphs like monopartite (homogenous type of users) and bipartite (two types of nodes for, e.g., users and posts), personal networks, groups, and pages.

These NASA export data in a common format, but to crawl the data from different sections of the FB network, we have to use some crawling algorithms. This lowers technical and logistical requests for empirical methods and helps researchers

Fig. 1 Various Network
Analysis Software
Applications (NASA) tools

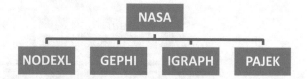

to explore an online medium that connects a billion of users treated as a walled garden. Many scholars have developed applications and tool-kits from sets of network analysis components, for example, the SMA (Statistics for Microarray Analysis) library, JUNG (Java Universal Network/Graph Framework) [21], GUESS (Graphical Unit Evolutionary Stochastic Search) [22], and Prefuse [23]. So, no need to develop a new NASA. There are other network analysis tools like NetDraw, UCINet, and Pajekc comprise rich graphical interfaces (GUI), complex and vast DLL's (Dynamic Link Libraries) of network metrics, and does not need technical skills like coding or command line execution typical features. The study discovers that these above-mentioned tools are developed for expert professionals, having complicated data handling, and not flexible enough for visualization and graph modeling features that retard wider acceptance.

In this chapter, we have selected a few tools for SNA for our stusdy considering in terms of platform, visualization features, file format, open source, language implementation, volume of input data, and other network properties like statistical measures, graph theory and so on. Network graph measures include diameter, density, centrality, the degree of hubs, betweenness measurement, and more. The different algorithms supported by these tools have incorporated the calculation of cyclic tree, weight of edges, etc. We have shortlisted few NASA from the Global Network for Social Network Analysis (INSNA) which keeps up a vast rundown of software applications, packages, and libraries. We have chosen for comparison namely NODE-XL, Gephi, IGraph,[8] Netvizz, and Pajek. The choices made are based on few facts. Each of the chose software is openly accessible to utilize and they can deal with a vast complex network. Almost all NASA is either in GUI format or libraries/packages which can be integrated as a part of a programming language. Gephi, Node-XL, Netvizz, and Pajek are GUI based NASA, though IGraph is a package-based application. Our aim is to explore various NASA in Fig. 1 that encourages bilateral analysis to uncover and investigate through "explicit" data manipulation, graphing and visualization. Following are the short insight about each of NASA tools.

4.1 Node-XL

This NASA application has a prepared and adaptable framework in Excel environment for Microsoft in 2007, 2010, and 2013 which finds the patterns easily and

---

[8]IGraph.sourceforge.net.

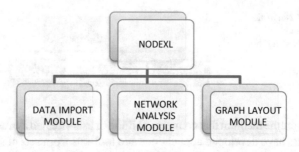

Fig. 2 Different phases of NodeXL modal

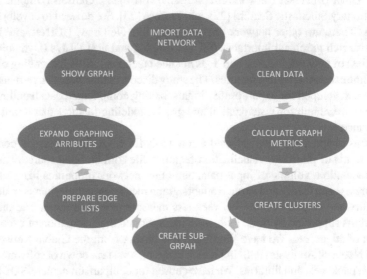

Fig. 3 Work flow model of the NodeXL-NASA

produces output in a well organize way in the format of graph layout. The few characteristic features of this product are: access to plug-in, community detection, Creating new edges based on closeness, computing statistical criteria, calculating various network metrics like eigenvector, clustering coefficient (CC), Page Rank, dynamic filtering, flexible and adaptable layout, zoom, direct fetching the network data from such as Facebook, email, twitter and other online social networks and expandable.

This section is an example of how to create your first Facebook network analysis. To use this, the first step is to create a basic visualization of a social media network. It is a free software application for excel 2007 and 2010 that lets the user enter a network edge list. Figure 2 depicts the various modules of this tool, i.e., system description In (Fig. 3), it shows the various steps to be followed to get network data visualization, i.e., workflow.

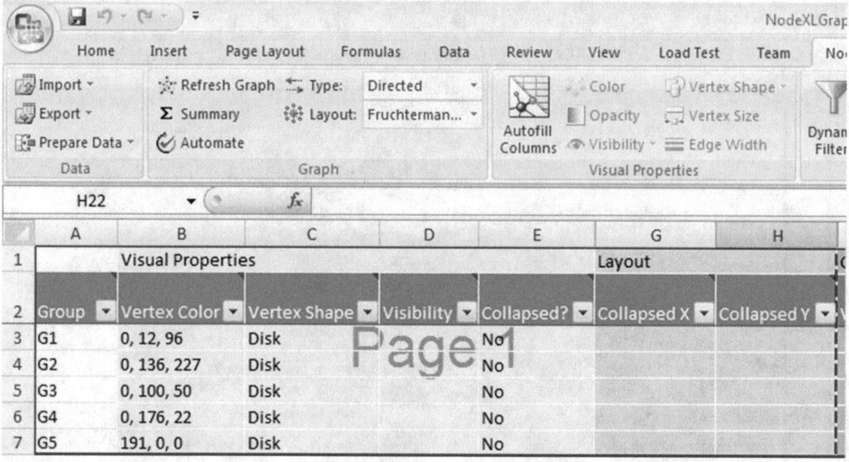

Fig. 4 NodeXL front-page

Figure 4 is the graphical user interface (GUI) of the NodeXL and Fig. 5 displays the "raw" network of the "friend" affiliations that members have created when they pledge and let one another explicit connection in the internal social network service application.

For the NodeXL, the data was from the Facebook network through a plug-in used for NASA-NodeXLp is "Social Network Importer". By using NASA, the network data statistics are generated as shown in Fig. 6, furthermore in Fig. 7, the figure shows those vertices have the maximum in-degree (red color), while remaining other vertices and connected links are gray color and the final picture of the social media network comprises the most linked (or connected) vertices in the online social media graph. In the same notion, in Fig. 6, the study retrieves the sub-images of the graph which are self-centered (egocentric) and can be infused in line with the different node data in the NODE-XL vertices worksheet.

This lets us sort all the sub-images in parallel with other data. Vertices can be arranged in any column or in any order by using the NODE-XL spreadsheet sort feature like in Fig. 7, the vertices can be placed by decreasing order of their clustering coefficient (CC). So, by making use of NASA NODE-XL, the study gets the visual representation of data and also able to fetch the statistical depiction and analysis stationed on the nontrivial observations. The other advantage of using this application on retrieved social media network dataset is that it allows us to find and highlight structural patterns like the different contributors, network key metrics, the number of connections within the network, etc. All these statistics and analysis operations by the NASA can be applied in a useful manner to the different social media data sets shared in status updates and can also be saved and copied to the local hard drive. The text information is machine readable.

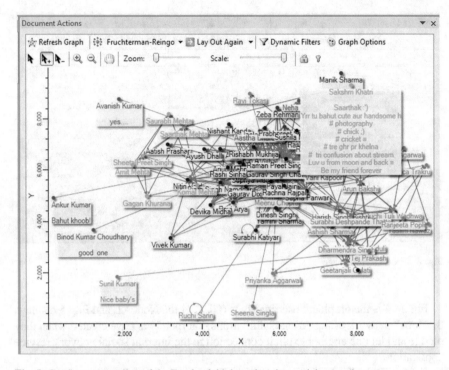

Fig. 5 Graph representation of the Facebook higher education social network

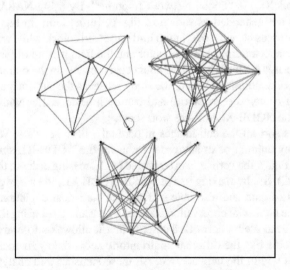

Fig. 6 NodeXL sub-graphic images of three users showing node networks

Fig. 7 Filtered network, highlighting the list of most connected nodes and they're the connections to one another

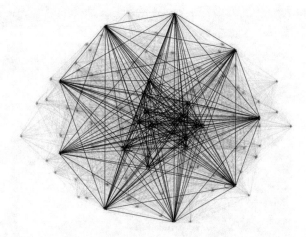

4.2 Analysis by GEPHI

The flexible GUI of Gephia (Figs. 8 and 9) is used to visualize, manipulate, and analyses networks. It is a powerful tool that has an active developer community and more efficient than the NASA-NODEXL in producing graphs. Gephi is an open-source network analysis application for presentation and visualization and helps researchers to show trending motifs or patterns related to isolated or outliers members and specifies other predictions from the data.

This application can calculate the network metrics like average diameter, clustering coefficients, etc., as shown in Figs. 10 and 11 and has a three-dimensional rendering engine to show large and complex real-time graphs in less time.

Gephi has a flexible architecture which combines all the built-in functionalities to explore analyses, specialize, filter, cluster, manipulate, and export many networks. This application is based on visualization and allows the user to alter the picture which allows any user to find networks and its data related properties. Gephi is designed in such a way; it follows the chain of a case study, from the data file to nice printable maps.

It is free software distributed under the GPL 3 (GNU General Public License). The effective module of this network analysis application is to work on the fetched network data file either from one of the network analysis application for e.g., Netvizz, or from any other external data sources, in the format of compatible graphics file (.Gdf) (as shown in Figs. 8, 9 and 11). While executing, the fetched data set or graph file will send all the data related to the network to the dynamic controller to display the visualization outcomes or output instantaneously and randomly.

This NASA is platform independent and the final data source file can be retrieved and easily compatible with other third party tools or applications (software's). The study explored the important features of this network exploration application, for e.g., annotation, clustering, data filtering, high-quality different layout algorithms

Fig. 8 Loading of the social media data set (Like personal network, a group data, any page)

and statistics. This application is user-friendly, scalable and flexible. Figures 12 and 13 display the high configured real-time layout of various algorithms available in the graph window.

Various properties like inertia, repulsion, and instance speed, size-adjust gravity, auto-stabilize and so on are available to figure the graph structure. There are other real-time settings available for the Force Atlas algorithm an important and vital force-directed algorithm present in NASA-Gephi. The GUI (graphical user interface) is also very streamlined or configured into different workspaces, where our study can do any isolated task and a strong and powerful application which can be plug-in onto any platform. It can be done for any end-user to add user defined utility like the filter or the algorithm. To filter out few vertices or links of the network, our study has to either do it manually or by using the filter system. This system filtered out few selected vertices or links with their network metrics like thresholds, clustering coefficient, range, radius, and other properties. To have a cluster of vertices or nodes, graphical user modules like Color Gradient, Size Gradient, color clusters, it can be possible to

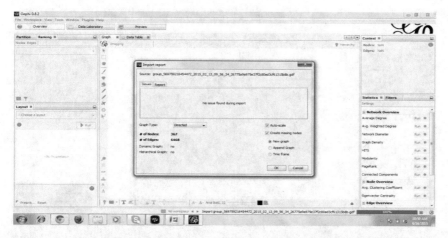

Fig. 9 Loading network data as nodes, vertices, directed or undirected (Connected graph for the group "Vivekananda School of Information Technology")

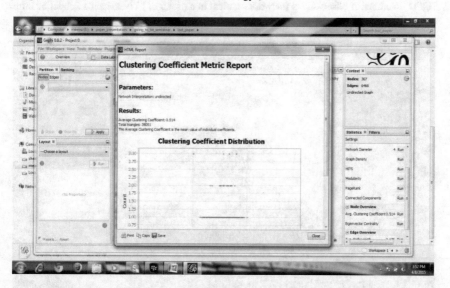

Fig. 10 Network metrics showing for e.g., clustering coefficient average diameter, no. of communities in a network, etc. (Connected graph for the group "Vivekananda School of Information Technology")

change the network visualization design. These different graphical modules take a set of vertices as the input parameter and alter the visualize parameters, like size or colors, to understand and authenticate the actual network framework or the content.

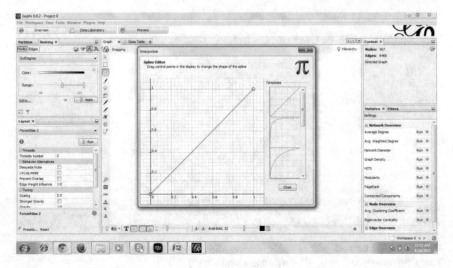

Fig. 11 Number of observations (network metrics) in a group of "Vivekananda School of Information Technology"

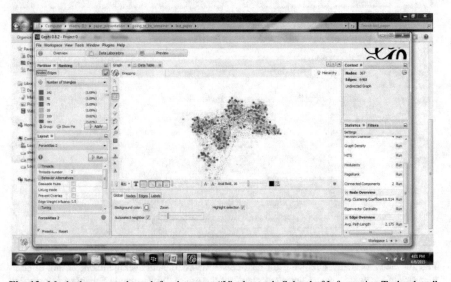

Fig. 12 Meshed connected graph for the group "Vivekananda School of Information Technology"

5 Comparison Between NASA App's

Social network analysis (SNA) tools like Gephi and Pajek are stand alone software applications, IGraph and Node-XL are the packages or libraries. Gephi, Pajek and Node-XL run on windows where IGraph uses python or R or C library for SNA.

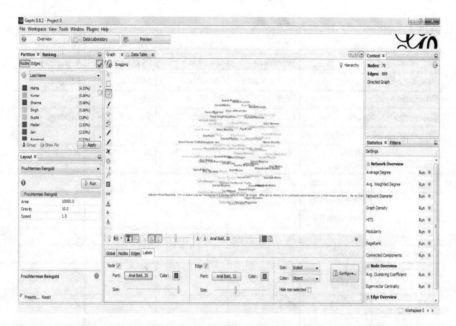

Fig. 13 Graph representation of the Facebook higher education (HE)—Vivekananda School of Information Technology—(VSIT)

Gephi can deal with up to 150,000 whereas IGraph, Node-XL or Pajek can deal with additional than one million nodes.

In SNA, we have many algorithms for layout. Pajek or IGraph have most recent and renowned algorithms like Kamanda Kawai or Fruchterman Reingold. All NASA tools have spring and circular layout. Gephi has more user-friendly graphical interface layout capabilities, for example, Photoshop features. Another famous algorithm supported by Gephi is Force layout.

Gephi is easy to handle and visualize small graphs, but not useful in case of complex and large graphs. For big and complex graphs we can use either IGraph or Node-XL and for small dataset we can use NASA tools like Pajek and Gephi.

In SNA, there are four categories of network graph. First, one-mode network, every node is related to each other node. In this type of network, we have just a single set of nodes and its ties are associated with these nodes. Second, a two-mode network, nodes are separated into two sets and nodes can link to another set. In multi-relational network there will be different sorts of relation's between nodes. For one-mode- or two-mode network analysis, we can use any of NASA tool, but for dynamic and growing graphs we can use Node-XL or Pajek. Comparisons among the four social networks on the basis of input file formats and graph features are shown in Table 1 respectively.

Table 1 Comparison between NASA tools

Parameter		Gephi	Pajek	IGraph	Node-XL
Platform used		Windows	Windows	Python	Windows
Number of users		Half million	Million	Million	Million
Processing time		Fast	Bit slow	Fast	Fast
Category of software		Library	Library	Stand alone	Library
Graph network types supported					
1	Multi-relational	✗	✓	✗	✗
2	Type-1 mode	✓	✓	✓	✓
3	Type-2 mode	✓	✓	✓	✓
4	Temporarily				
		✗	✓	✗	✓
Graph layout supported					
1	Fruchterman reingold	✗	✓	✓	✗
2	Random layout	✗	✓	✓	✓
3	Force altas	✗	✓	✗	✗
4	Graphviz	✗	✗	✗	✓
5	Spectral	✗	✗	✗	✓
6	Circular	✓	✓	✓	✓
7	Spring layout	✓	✓	✓	✓
File format supported					
1	.DAT	✗	✓	✗	✗
2	.Pajek	✗	✓	✓	✗
3	.GDF	✓	✗	✗	✗
4	.GML	✓	✗	✓	✓
5	.CSV (EdgeList)	✗	✗	✓	✓
6	.TXT (Edgelist)	✗	✗	✓	✓
7	.GraphGML	✓	✗	✓	✓
8	.NET format	✓	✓	✓	✓
9	Graph DB	✓	✗	✓	✗
10	Adjacency list	✗	✗	✓	✓

6   Case Study "Vivekananda School of Information Technology" (VSIT)

Our approach is to use NASA for retrieving all friendship, relationship connections between the members of the group. Our study found out a few things about the group VSIT: (i) High "network density" 26% of all possible ties (i.e., Graph density is 0.26), (ii) an average degree of 39.7 (each node is visualized within its own "1.5 degree" network of associations to and from the node itself) which shows that group holds a tight bonding with each other rather than lose coupled (attached for sharing information about on any debatable subject), (iii) the patterns predicted from the visualization of the graph states that the relationships were not dispersed, (iv) members (approx 2.2%) of the group are isolated from others, i.e., no friendship pattern exists for them—this predicts that the sample size of isolation, population is interested in the current subject rather than through social ties-up, (v) 47.8% have at almost 30 connections and 22.5% have more than 80 connections. By viewing the number of connections may consider as one way to show the group leader, social media network analysis provides a massive arsenal of methods to analyze and visualize graphs in more specific ways. In Figs. 12 and 13 study has shown the spatial visualization of the group by using NASA-Gephi and analyzed the dataset provided by NASA-Netvizz. The different network measurements are calculated like betweenness centrality (BC), the topology of graphs, etc. These network metrics help us in detecting a node's position in the larger complex social media graph and is useful in measuring strategic positioning rather than based on social status or popularity index. A node or user having high BC is treated as the most influential person (to distort or withholding the information) in the group because of his location in the network. The high BC denoted the concept of bridging-capacity to coupled or gel separate groups. In our case study, group creator or administrator is an overall central figure or Bridger which has high BC.

Almost all the NASA tools offer implementations of these network metrics (e.g., BC) for researchers who can investigate the graph structure disintegrated into groups and so on. Other metrics which the study has explored are the user interface "locale" that is available for each user on Facebook. At last, we can conclude from this study that though this group has a high level of connectivity yet it is retained its national coherence.

7   Conclusion and Summary

The purpose of this chapter is to discuss various Network analysis software applications for the different subsections of the social media Facebook network. Presently, Facebook has over one billion online users, so it is crucial to create and to solidify research models for the ongoing communication networks. The study uses NASA for the empirical research purpose only without entering into contractual partnerships with the various IT companies. All these above discussed NASA can be developed further and as a future scope, they may be upgraded for handling temporal data mining. The study made two conclusions about NASA's these are: (i) All of NASA's are interoperable and the data sources fetched can easily adapt and communicate with the other third party vendors, databases or web services or existing software's, and (ii) NASA's generates such outcomes whose structure or framework or content evolves with time, and suggest a time bound constituent where a part of the social media network can be fetched.

NASA stand alone tool is very valuable for network Visualization—up to a couple of thousands of nodes and also useful for data conversions. IGraph package which we used in R programming is fastest tool among all NASA app's that display the vast majority of graph properties, features and handle huge and complex graph network. Libraries like IGraph or Node-XL are more helpful for tasks having a huge number of nodes and for operations like the difference and the union of sets of nodes or for the grouping (clustering). Standalone applications are easy and simple to learn, but for beginner level Pajek and Gephi is appropriate applications. Research scholars can use IGraph programming and Node-XL for their real-world vast and complex datasets.

Network analysis in one mode or two mode we can use any of NASA however for multi-social (relational) network graph, we have only one, i.e., Pajek application. We can use Node-XL and Pajek application for temporarily network.

Almost all the NASA tools can handle data files in format of .Net. But in many of the cases, data are available in .txt format. We can easily handle .txt format or number of SNS gives data in .txt format so we can utilize IGraph or Node-XL applications. Many online applications or tools are available to convert .txt format into .Net format but these few NASA can deal with just small size file. So, therefore, we can use Node-XL or IGraph applications for the large data sets.

IGraph package delivers mostly all graph features and it additionally it can handle complex and large network. Every one of NASA can process density, network diameter, centrality, page rank, clustering coefficient (CC). But to compute some specific features like the Cohesion then we have to choose IGraph or Node-XL applications. Similarly, if we have to compute Dyad and Bridge, we can use Pajek, IGraph, or Node-XL applications where as Gephi does not give any calculation related to dyad or bridge.

Pajek and IGraph both are faster NASA Tools with respect to others. But drawback of Pajek is that it does not display all graph highlights or features. So to examine all the graph features and network properties we can utilize IGraph programming. IGraph produces a quickest outcome to almost all the network metrics. Loading time is least for Pajek and execution time for all network metrics like CC, average diameter, graph nodes, cliques is least in IGraph. So, IGraph is better in producing faster results.

Network data structures are crucial in almost all the professions and research areas. The interesting part in these network data structures is that they are growing more common as the world's real online social media networks grow. These NASA tools aim to do analysis and visualization of online social media network data easily by integrating the visualization functions and common analysis modules with the each other paradigms for data handling. These NASA tools not only enable us to do an effective network analysis tasks, but also supports a large target audience of users in a wider spectrum of network analysis scenarios. Using these software applications on a sample social media network dataset like "Higher Education in India" the study fetched and shown the framework patterns like the intensity, density of users' connections within the network or group, different contributors, and key network metrics. These analyses and visualization tasks can be usefully applied to a broad range of online social media networks. This chapter provides an overview of analytical directions explored so that the data made accessible, considers platform-specific aspects of data extraction via the official Application Programming Interface (API), and concisely engages the crucial ethical considerations attached to this type of research.

References

1. Mislove, A., Marcon, M., Gummadi, K.P.: Measurement and Analysis of Online Social Networks. Max Planck Institute for software Systems
2. Carrington, J.P., Scott, J.: For a historical overview of the development of social network analysis. Introduction. The Sage Handbook of Social Network Analysis, p. 1. SAGE (2011). ISBN 978-1-84787-s395-8
3. Boyd, D.M., Ellison, N.B.: Social network sites: definition, history and scholarship. J. Comput. Mediat. Commun. **13**(1), 210–230 (2008)
4. Nickson, C.: The history of social networking. Digital trends (2009). http://www.digitaltrends.com/features/thehistory-of-social-networking. Accessed 17 Feb 2010
5. Dwyer, C., Hiltz, S.R., Passerini, K.: Trust and privacy concern within social networking sites: a comparison of Facebook and myspace. In: 13th Americas Conference on Information Systems, Colorado, USA, 9–12 Aug 2007
6. Douglis, F.: It's all about the (Social) network. All system go. J. IEEE Internet Comput. **14**(1): 4–6
7. Qiu, J., Lin, Z., Tang, C., Qiao, S.: Discovering organizational structure in dynamic social network. In: 9th IEEE International Conference on Data Mining, pp. 932–937, Miami, Florida, USA, 6–9 Dec 2009
8. Monclar, R.S., et al.: Using social networks analysis for collaboration and team formation identification. In: 2011 15th International Conference on Computer Supported Cooperative Work in Design (CSCWD). IEEE (2011)
9. Akhtar, N., Javed, H., Sengar, G.: Analysis of Facebook social network. In: IEEE International Conference on Computational Intelligence and Computer Networks (CICN), Mathura, India, 27–29 Sept 2013
10. Zelenkauskaite, A., et al.: Interconnectedness of complex systems of internet of things through social network analysis for disaster management. In: 2012 4th International Conference on Intelligent Networking and Collaborative Systems (INCoS). IEEE (2012)
11. Li, J., Chen, Y., Lin, Y.: Research on traffic layout based on social network analysis. In: 2010 2nd International Conference on Education Technology and Computer (ICETC), vol. 1. IEEE (2010)
12. Dekker, A.: Conceptual distance in social network analysis. J. Soc. Struct. **6**(3), 1–34 (2005)
13. Jensen, D., Neville, J.: Data mining in social networks. In: National Academy of Sciences Symposium on Dynamic Social Network Modeling and Analysis, pp. 287–302, Washington D.C., USA, 7–9 Nov 2002

14. DeRosa, M.: Data mining and data analysis for counterterrorism. CSIS Report. The CSIS Press (2004)
15. Meenu, C., Mamta, M.: The education gets the facelift by going social. Int. J. Appl. Innovation Eng. Manag. (IJAIEM) 2(12), 50–53 (2013). ISSN 2319-4847
16. Mamta, M., Meenu, D., Meenu, C.: Social network analysis (SNA). In Facebook higher education groups through NASA (Network analysis software applications). Int. J. Artif. Intell. Knowl. Discov. (IJAIKD). ISSN 2231–0312
17. Batagelj, A.M.V.: Pajek—program for large network analysis. Connections 21(2), 47–57 (1998)
18. Freeman, L.C., Borgatti, S., Everett, M.G.: UCINET (6). Analytic Technologies (2006)
19. Backstrom, L., Marlow, C., Ugander, J., Karrer, B.: The anatomy of the Facebook social graph (2011). arXiv:1111.4503
20. Wimmer, A., Gonzalez, M., Lewis, K., Christakis, N., Kaufman, J.: Tastes, ties, and time: a new social network dataset using facebook.com. Soc. Netw. 30(4), 330–334 (2008)
21. Whit, S., O'Madadhai, J., Smyth, P., Boey, Y.B., Fisher, D.: Analysis and visualization of network data using JUNG. J. Stat. Softw. VV (2005)
22. Adar, E.: GUESS: a language and interface for graph exploration. In: Proceedings of ACM Conference on Human Factors in Computing Systems (2006)
23. Landay, J., Card, S.K., Heer, J.: Prefuse: a toolkit for interactive information visualization. In: Proceedings of ACM Conference on Human Factors in Computing Systems (2005)

Machine Learning on Big Data: A Developmental Approach on Societal Applications

Le Hoang Son, Hrudaya Kumar Tripathy, Biswa Ranjan Acharya, Raghvendra Kumar and Jyotir Moy Chatterjee

Abstract Machine Learning (ML) is a potential tool that can be used to make predictions on the future based on the past history data. It constructs a model from input examples to make data-driven predictions or decisions. The growing concept "Big Data" need to be brought a great deal accomplishment in the field from claiming data science. It gives data quantifiability in a variety of ways that endow into data science. ML techniques have made huge societal effects in extensive varieties of applications. Effective and interactive ML relies on the design of novel interactive and collaborative techniques based on an understanding of end-user capabilities, behaviors, and necessities. ML could additionally make utilized within conjunction for enormous information to build effective predictive frameworks or to solve complex data analytic societal problems. In this chapter, we concentrate on the most recent progress over researches with respect to machine learning for big data analytic and different techniques in the context of modern computing environments for various societal applications. Specifically, our aim is to investigate opportunities and challenges of

The original version of this chapter was revised: Co-author name has been changed. The erratum to this chapter is available at https://doi.org/10.1007/978-981-13-0550-4_12

L. H. Son (✉)
VNU University of Science, Vietnam National University, 334 Nguyen Trai,
Thanh Xuan, Hanoi, Vietnam
e-mail: sonlh@vnu.edu.vn

H. K. Tripathy · B. R. Acharya
KIIT University, School of Computer Engineering, Patia 751024, Odisha, India
e-mail: hrudayakumar@gmail.com

B. R. Acharya
e-mail: acharya.biswa85@gmail.com

R. Kumar
LNCT Group of College, Department of Computer Science and Engineering, Jabalpur, India
e-mail: raghvendraagrawal7@gmail.com

J. M. Chatterjee
Department of Computer Science and Engineering, GD-RCET, Bhilai 490024, CG, India
e-mail: jyotirm4@gmail.com

© Springer Nature Singapore Pte Ltd. 2019
M. Mittal et al. (eds.), *Big Data Processing Using Spark in Cloud*,
Studies in Big Data 43, https://doi.org/10.1007/978-981-13-0550-4_7

ML on big data and how it affects the society. The chapter covers discussion on ML in Big Data in specific societal areas.

Keywords Machine learning · Big Data · Societal application · Big data analysis

1 Introduction

Machine Learning (ML) is a kind of tool used for predictions based on past reference data to make data-driven predictions or decisions for the future. The healthy concept on "Big Data" has successfully used in the field of data science. Building effective predictive systems or to solve complex problems through data analytic, ML plays a major role in coincidence with Big Data [1]. According to a present study, the use of machine learning algorithms replaces high percentage of employment across the globe in future. The applications of machine learning algorithms are highly self-modifying and automated model to learn with sufficient data which helps to provide the improvement of over time with minimal human intervention [2]. The present big data scenarios are rapidly spreading in different domains of science as well as in engineering area. Undoubtedly, the usability of such massive data fully makes sense of new way thinking and supports to create new techniques of acquisition to face different challenges. In this chapter, we present a literary study on machine learning techniques for Big Data and highlight some promising learning methods. Also, we focus on different kind of techniques and discussions on possible applications of machine learning in Big Data for societal utilization. The rest of this chapter is organized as follows. Sections 2 and 3 present classification of machine learning algorithms. Details of Regression, Decision Tree, Support Vector Machine, Naïve Bayes, Artificial Neural Network, k-Nearest Neighbor, Dimensionality Reduction, Gradient Boost and Adaboost are introduced in subsequent sections. Lastly, conclusions are highlighted in the last section (Fig. 1).

2 Classification of Machine Learning Algorithms

Machine learning is an important area of artificial intelligence. The objective of machine learning is to discover knowledge and make intelligent decisions. Machine learning algorithms basically grouped into supervised, semi-supervised, and unsupervised. When big data are concerned, it is necessary to scale-up machine learning algorithms [1, 2]. Another categorization of machine learning according to the output of a machine learning system includes classification, regression, clustering, and density estimation, etc. Machine learning overcomes the limitations that are impacted by the human capability of thinking and finding correlations in huge amounts of data. Machine learning systems are able to explore and display patterns hidden in such data. They use automated methods for big data analysis. The intention of delivering accurate predictions it needs the direction for the improvement of fast, cost effective, and efficient algorithms in real-time processing of data. Normally, big data are large volume of both structured and unstructured data that and it is obviously difficult to process by taking the help of database concept and traditional software techniques. Machine learning is helping in great way to big data and gives an impact on technological discoveries and value addition [3, 4].

2.1 Supervised Learning Algorithms

Machine learning methods are diverse sort of models that make forecasts on set of tests gave. In the event of directed learning calculation, input data is called training data which looks at for changed sort of examples inside the esteem depiction appointed to various data focuses. Through a training procedure a model is readied where it is required to make expectations and will make revision when those forecasts are not right. This strategy contains a target/result variable (or ward variable) which is to be foreseen from a given course of action of indicators (self-governing factors). Using these game plan of elements, we make a limit that guide commitments to desired yields.

2.2 Unsupervised Machine Learning Algorithms

In unsupervised learning, data is not named and does not have a known result. Machine learning systems build the data into a surge of clusters and there are no names related with information centers. A model is set up by intuition structures display in the information.

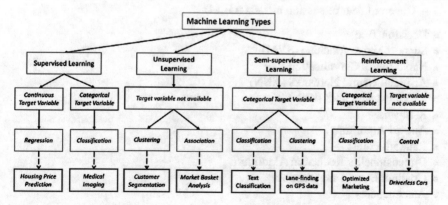

Fig. 1 Type of machine learning

2.3 Semi-supervised Learning

On the off chance that there ought to emerge an event of semi-supervised learning, the data are blend of checked and unlabeled data. It can have demand of prediction issue that must take in the structures to sort out the data and make desires. Portrayal issues like categorization and relapse [5].

2.4 Reinforcement Machine Learning Algorithms

This is a range of machine learning motivated by behaviorist brain research, worried about how programming operators should take activities in a situation in order to augment some idea of aggregate reward. Fortification learning contrasts from standard administered learning in that right info/yield sets are never introduced, nor imperfect activities unequivocally remedied [6].

3 Machine Learning Algorithms

These are the following mentioned list of generally used machine learning methods. It can be used to solve almost any real applications:

- Regression
  - Linear Regression
  - Logistic Regression
  - Ordinary Least Squares Regression
  - Stepwise Regression
  - Multivariate Adaptive Regression Splines (MARS)
  - General linear regression models (GLM)

- Decision Tree
- Support Vector Machine (SVM)
- Naïve Bayes Classification
- Artificial Neural Networks (ANN)
- k-Nearest Neighbor (kNN)
- K-Means
- Apriori Algorithm
- Random Forest
- Dimensionality Reduction Algorithms
- Gradient Boost and Adaboost
- Autoencoders

While crunching information to prime example business choice, people normally using most commonly supervised and unsupervised learning techniques. A fascinating issue in the present time is semi-supervised learning techniques fields, for example, picture categorization where there are expansive datasets with a modest number of named outlines [7]. There are diverse ways an algorithm can demonstrate an issue in light of its association with the experience or condition or whatever we need to call the data. There are just a couple of fundamental learning methods or models that a technique can have well experience to make solution for different problem types that they suit. This scientific categorization or method for sorting out machine learning calculations is helpful in light of the fact that it compels you to consider the parts of the information and model readiness process and select the ones that is having the most fitting for ones concern keeping in mind the end goal to get the best outcome.

4 Regressions

These are worried about demonstrating the connections between variable which can be constantly cultivated utilizing a part of fasten up the forecast built utilizing the model. Regression techniques were strategies for measurements that can be used into quantifiable machine learning. It might be baffling out of the way that can utilize regression to intimate the level of issue and class of strategy [8]. Genuinely, regression is a methodology. Some representation methods are:

4.1 Linear Regression

They are prominent Machine Learning Algorithm, and the most utilized one today. It takes a shot at continuous factors to make forecast. Linear Regression endeavors to shape a connection among autonomous and subordinate factors and to frame a regression line, i.e., a "best fit" line, used to make future forecast. The motivation behind this regression line is to lower the separation between the data indicates and the regression line makes a mathematical expression as

$$Y = a * X + b$$

i.e., the expression of a straight line where, Y is the dependent variable, X is the independent variable, as in slope, and b is the intercept. The premise of the model depends on the least squares calculation approach, among other minimum utilized methodologies. There are two kinds of Linear Regressions (Fig. 2):

- **Linear Regression**—where there is just a single independent variable.
- **Multiple Linear Regression**—where there is more than one independent variable.

Fig. 2 Linear regression

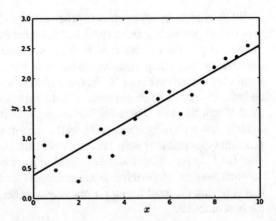

4.1.1 Benefits of Linear Regression Machine Learning Algorithm

- It is a standout among the most accountable machine learning methods, making it simple to describe to others.
- It is simple to utilize as it requires negligible adjustments.
- It is the most broadly utilized machine learning procedure that runs quick.

4.1.2 Uses to Mankind of Linear Regression

- **Estimating Sales**

Linear Regression discovers extraordinary use in business, for deals anticipating in light of the current direction. On the off chance that an organization watches relentless increment in deals each month—a linear regression evaluates of the month to month deals data enables the organization to forecast deals in up and coming months.

- **Risk Assessment**

Linear Regression evaluates chance associated with protection or budgetary area. A medical insurance coverage organization commits linear regression evaluation on the quantity of cases per client against age. This investigation helps insurance agencies locate, that more old clients tend to make more protection claims. Such investigations come about assume a crucial part in critical business decision and are made to represent risk.

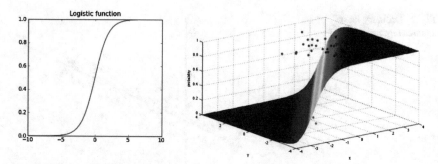

Fig. 3 Logistic regression

4.2 Logistic Regression

This is a Machine Learning algorithm where the reliant variable is unmitigated. It calculates just discrete esteems, similar to 0 or 1, yes or no. Evaluating probabilities utilizing a Logistic Function decides the connection between the dependent variable and the independent variables. The curve plotted between the variables makes a S-molded bend instead of a straight line in direct regression. Logistic Regression is utilized when the result to be anticipated binary—i.e., 0 or 1. Something else, different strategies like Linear Regression are picked [9] (Fig. 3).

Logistic regression is a capable statistical method for demonstrating a binomial result with at least one illustrative factor. It gauges the connection between the categorical reliant variable and at least one autonomous variables by assessing probabilities utilizing a logistic function, which is the aggregate logistic distribution. The name of this method could be a touch of overwhelming as in Logistic Regression method can be used in categorization assignments and not in regression issues.

4.2.1 Classification of Logistic Regression

- **Binary Logistic Regression**—The most normally utilized logistic regression when the categorical reaction has two conceivable results, i.e., either yes or not. Illustration—Predicting whether a learner will succeed or not an exam, anticipating whether a man will have low or hypertension, foreseeing whether a tumor is destructive or not.
- **Multi-nominal Logistic Regression**—All out reaction has at least three or more conceivable results with no requesting. Illustration predicting what sort of web search tool (Yahoo, Bing, Google, and MSN) is utilized by lion's share of US residents.
- **Ordinal Logistic Regression**—Ordinal Logistic Regression—Categorical reaction has at least three conceivable results with natural ordering. Illustration How a client rates the administration and quality of food at a hotel in view of a size of 1–10.

Fig. 4 Ordinary least
squares regression

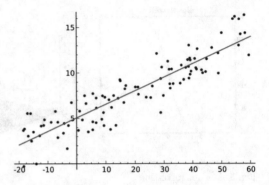

4.2.2 At the Point When to Utilize Logistic Regression Machine Learning Algorithm

Utilize logistic regression methods when there is an essential to foresee the probabilities of the straight out ward variable as a component of some other illustrative variable. For example, likelihood of acquiring a thing X as a segment of sexual introduction. It can be utilized when there is a need to anticipate probabilities that obvious ward variable will fall into two orders of the twofold response as an element of a few logical factors. It is also most appropriate when the need is to classify components two classifications in view of the illustrative variable. For instance characterize females into "youth" or "old" gathering in light of their age.

4.3 Ordinary Least Squares Regression

On the off chance that you know statistics, you most likely have known about linear regression some time recently. Least squares are a method for performing direct regression. You can consider direct regression as the task of fitting a straight line through a course of action of focuses. There are different possible systems to do this, and "ordinary least squares" strategy go like this — You can draw a line, and after that for each of the information centers, measure the vertical partition between the point and the line, and incorporate these up; the fitted line would be the one where this whole of separations is as little as could reasonably be expected (Fig. 4).

4.4 Stepwise Regression

It is a semi-robotized procedure of building a model by progressively including or excluding factors construct exclusively in light of the t-insights of their assessed coefficients. Appropriately utilized, the stepwise regression alternative in Stat graphics (or other stat bundles) puts more power and data readily available than does the normal numerous regression choice, and it is particularly helpful for filtering through huge quantities of potential independent factors and additionally tweaking a model by jabbing factors in or out. Disgracefully utilized, it might meet on a poor model while giving you an incorrect feeling that all is well with the world [6]. Stepwise regression is a way to deal with choosing a subset of impacts for a regression model. It can be valuable in the accompanying circumstances

- There is little hypothesis to control the determination of terms for a model.
- You want to intelligently investigate which prediction appear to give a exact match.
- It needs to enhance a models forecast execution by diminishing the difference caused by assessing pointless terms.

For categorical predictors, you can do the accompanying

- Choose from among different standards to decide how related terms enter the model.
- Enforce impact heredity.

The stepwise platform additionally empowers you to investigate every single conceivable model and to direct model averaging.

4.5 Multivariate Adaptive Regression Splines (MARS)

It was introduced by Friedman amid 1991. It has an adaptable procedure to oversee connections between an arrangement of info factors and the objective ward that are almost added substance or include cooperation's with fewer factors. It is a non-parametric measurable technique in view of a gap and vanquishes procedure in which the preparation informational indexes are apportioned into isolated piece insightful straight fragments (splines) of varying inclinations (slant). MARS makes no suppositions about the hidden utilitarian relationships between dependent and independent variables. In general, the spines are connected smoothly together, and these piecewise curves (polynomials), also known as basic functions (BFs), result in a flexible model that can handle both linear and nonlinear behavior. The connection/interface points between the pieces are called knots. Marking the end of one area of information and the start of another, the applicant ties are put aimlessly positions inside the scope of each information variable [10]. MARS generates BFs by stepwise searching overall possible unify variety candidate knots and across interactions among all variables. A versatile regression method is embraced for naturally choosing the bunch areas. The MARS calculation includes a forward stage and a retrogressive stage. The forward stage places hopeful bunches indiscriminately positions inside the scope of every indicator variable to characterize a couple of BFs. At each progression, the model adjusts the bunch and its comparing pair of BFs to give the most extreme decrease in whole of-squares leftover blunder. This procedure of including BFs proceeds until the point when the most extreme number is come to, which usually results in a very complicated and over fitted model. The backward phase involves deleting the redundant BFs that made the least contributions. An open MARS source code from Jekabsons (2010) is adopted in performing the analyses presented in this chapter.

Assume y a chance to be the objective ward reactions and $X = (X_1...X_p)$ be a network of P input factors. At that point it is expected the information are created in view of an obscure "true" model. For a consistent reaction, this would be

$$Y = f(X_1 ... X_p) + e = f(X) + e \tag{1}$$

In which e is the fitting error is the built MARS model, comprising of BFs which are spines piecewise polynomial functions. For simplicity, only the piecewise linear function is expressed and considered in this chapter. Piecewise linear functions follow the form max(0, x–t) with a knot defined at value t. Expression max(.) means that only the positive part of (.) is used otherwise it is assigned a zero value. Formally,

$$\text{Max}(0, \ \text{x} - t) \begin{cases} x - t, & if \ x \geq 0 \\ 0, & otherwise \end{cases} \tag{2}$$

The MARS model f(X), which is a linear combination of BFs and their interactions, is expressed as

$$f(x) = \beta 0 + \sum_{m=1}^{M} \beta m \lambda m(X), \tag{3}$$

where each λm is a BF. It could be a spline function, or interaction BFs generated by multiplying current term with a truncated linear function including another/diverse variable (higher requests can be utilized just when the information warrants it; for straightforwardness, at most second request is received). The term β is constant coefficients, estimated using the least-squares method.

4.6   General Linear Regression Models (GLM)

An extremely basic approach in light of determining a linear mix of predictors to predict a reliant variable. Coefficients of the model are a measure of the quality of impact of every indicator on the result. It incorporate direct and logistic regression models can present over fitting and multidisciplinary in high-dimensional problems

4.6.1   Applications of Regression Algorithm on Big Data

- Machine learning, statistical learning and the future of biological research in psychiatry—Psychiatry has now gone to level of research where techniques of big data are of significant importance. General linear regression models (GLM) is used to predict a dependent variable for other machine learning algorithms.

- Big Data Analysis Using Modern Statistical and Machine Learning Methods in Medicine—Nowadays present day measurable machine learning and bioinformatics approaches are utilized as a part of learning factual connections from enormous information in pharmaceutical and socio-relational science that ordinarily incorporate analytical, genetic (and proteomic), and ecological factors. Direct and sensible regressions are generally utilized as a part of clinical information examinations. In the field of clinical information, direct regression demonstrate with settled impacts were connected at tolerant level and it has been discovered that among patients with diabetes, the utilization of an EHR was related with an unassuming lessening in ED visits and hospitalizations yet not on office visit zones. Legitimate regression is broadly utilized as a part of result variable that has two results, e.g., regardless of whether a patient is experiencing a specific sickness or not.
- Automated Trading with Machine Learning on Big Data—This is to produce benefit from different between showcase value forecasts and markets' relationship structure. Highlight extraction is finished utilizing exogenous exchange flag era process where a distinction between the objective esteem and its predictor's saw desire can be drawn from the basic straight regression fitted over the look back.
- A Taxonomy of Big Data for Optimal Predictive Machine Learning and Data Mining—Statistical machine learning methods used for massive data uses logistic regression beside some other algorithms for calculating the sample size and classification.

5 Decision Trees

A decision tree is a choice help instrument that uses a tree-like graph or model of choice and their possible outcomes, including chance-event comes about, resource costs, and utility. Examine the photo to get a sentiment what no doubt like. Decision Tree is a champion among the most well known managed Machine Learning methods. It is moreover the most easy to get it. Decision trees basically perform categorization, however normally, it shapes regression trees too. They work for straight out and nonstop info and yield factors. A decision tree parts the information into packs that are homogeneous yet heterogeneous to each other. The tree makes utilization of an assortment of strategies to settle on this decision. Be that as it may, the objective of the technique is this—Splits are performed on all criteria, and the split that outcomes in the most homogeneous subgroup are picked [8] (Fig. 5).

From a business decision viewpoint, a decision tree is the base number of yes/no request that one needs to request, to overview the probability from settling on a correct choice, as a rule. As a method, it empowers you to approach the issue in a sorted out and think way to deal with arrive at a predictable conclusion. A choice tree is a graphical depiction that makes use of extending method to epitomize each and every possible aftereffect of a choice, in perspective of particular conditions. In

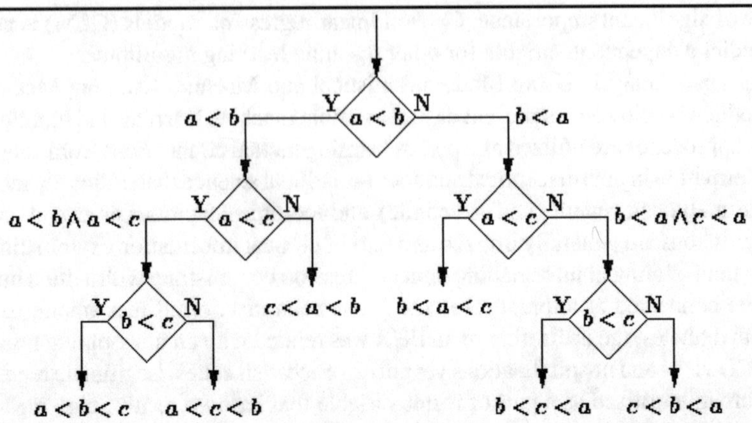

Fig. 5 Decision trees

a decision tree, within center point/hub addresses a test on the quality, each branch of the tree addresses the aftereffect of the test and the leaf center point/hub addresses a particular class check, i.e., the decision made in the wake of figuring most of the qualities. The gathering rules are addressed through the path from root to the leaf hub.

6 Support Vector Machine (SVM)

Every individual information points organize in the dataset is a Support Vector. The Support Vectors in the wake of plotting outline a characterization. SVM is parallel characterization method. Given a course of action of purposes of 2 sort of in N dimensional place, SVM makes a $(N - 1)$ dimensional hyper-path to isolate those focuses into 2 clusters. We should expect you have a couple of motivations behind 2 sorts in chapter which are directly separable. SVM will find a straight line which disengages those focuses into two sorts and organized past what many would consider conceivable from each one of those focuses. As far as scale, a portion of the most serious issues that have been illuminated utilizing SVMs (with appropriately changed executions) are show publicizing, human graft site recognition, picture based gender detection, vast scale picture classification. Bolster Vector Machine, administered Machine Learning algorithms, utilized for classifications and regression problems, however generally more for classification. It makes for an exceptionally viable instrument, utilized essentially while classifying a thing into one of two groups [11] (Fig. 6).

Kernel Methods are best known for the notable procedure Support Vector Machines which is genuinely a star gathering of systems without anyone else's input. Kernel Methods are ready about mapping input information into a higher dimensional vector space where some grouping or regression issues are most effortless for demonstrating [8].

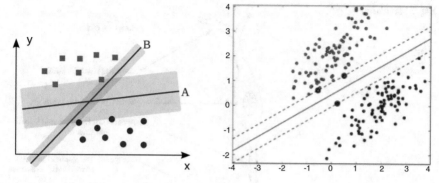

Fig. 6 Support vector machine

7 Naïve Bayes Classification

Naive Bayes classifiers are a group of straightforward probabilistic classifiers in view of applying Bayes' hypothesis with solid (innocent) autonomy assumptions between the features. The included image is the equation — with P(A|B) is back likelihood, P(B|A) is probability, P(A) is class earlier likelihood, and P(B) is predictor earlier probability [4]. Some of genuine cases are:

- To check an email as spam or not spam
- Group a news article about innovation, legislative issues, or games
- Check a bit of content communicating positive feelings, or negative feelings?
- Utilized for face recognition shareware.

8 Artificial Neural Networks (ANN)

Artificial Neural Networks are moderately rough electronic models in light of the neural structure of the brain. The brain essentially gains from observation of fact. It is normal confirmation that a few issues that are past the extent of current computer systems are to be sure feasible by little vitality effective bundles. This cerebrum modeling additionally guarantees a less specialized approach to create machine solutions. This new way to deal with execution likewise gives a more smooth debasement amid framework over-burden than its more customary partners. These biologically roused techniques for calculating are believed to be the following real headway in the evaluating business and industry. Indeed, even straightforward animal brains are equipped for capacities that are as of now incomprehensible for computers. Computer Systems do repetition things well, such as keeping records or performing complex math. Be that as it may, computers experience difficulty perceiving even straightforward patterns considerably less summing up those patterns of the past into activities without bounds [12].

Fig. 7 A neuron

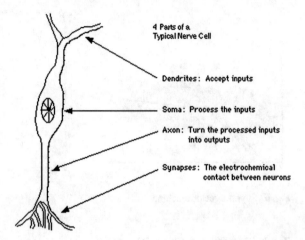

Presently, propels in biological research guarantee an underlying comprehension of the normal thinking technique [13]. This research exhibits that brains store data as examples. Some of these examples are to a great degree obfuscated and empower us the ability to see singular appearances from an extensive variety of edges. This method of securing data as examples, utilizing those examples, and subsequently dealing with issues incorporates another field in assessment. This field, as said some time as of late, does not utilize customary programming yet rather incorporates the creation of hugely parallel systems and the preparation of those frameworks to deal with specific issues. These fields moreover utilizes words through and through not the same as customary assessing, words like act, react, self-mastermind, learn, sum up, and neglect [7]. Artificial neural networks (ANNs) are computational models enlivened by the human mind. At their core, they are comprised of a large number of connected nodes, each of which performs a simple mathematical operation. Each node's output is determined by this operation, as well as a set of parameters that are specific to that node. By connecting these nodes together and carefully setting their parameters, very complex functions can be learned and calculated. Artificial neural networks are responsible for many of the recent advances in artificial intelligence, including voice recognition, image recognition, and robotics. For example, ANNs can perform image recognition on hand drawn digits [3]. The individual neurons are convoluted. They have a stack of parts, sub-structures, and control frameworks. They pass on data by methods for an expansive gathering of electrochemical pathways. There are more than one hundred unique classes of neurons, contingent upon the order strategy utilized. Together these neurons and their associations shape a procedure which is not parallel, not steady, and not synchronous. To put it plainly, it is not at all like the right now accessible electronic PCs, or even simulated neural systems (Fig. 7).

Fig. 8 Artificial neuron

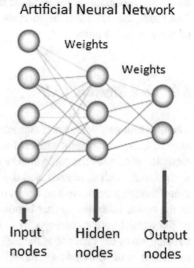

The input nodes take in data, in the frame which can be numerically communicated. The data is represented as activation values, where every node is given a number, the higher the number, the more prominent the activation. This data is then passed all through the system. In light of the association values (weights), restraint or excitation, and transfer function, the activation value is passed from node to node. Each of the nodes totals the activation values it gets; it at that point changes the value in view of its transfer function. The activation courses through the network, through hidden layers, until the point that it achieves the output nodes. The yield nodes at that point mirror the contribution to an important route to the outside world. The distinction between anticipated value and real value (error) will be proliferated in reverse by distributing them to every node's weights as per the measure of this error the node is in charge of. Neurons are associated with each other, with every neuron's

8.1 Mechanism of Artificial Neuron

The ANN endeavors to reproduce the computational reflection of the natural neural system, despite the fact that it is not tantamount since the number and multifaceted nature of neurons and the utilized as a part of a biological neural network is ordinarily more than those in an artificial neural network. An ANN is included a network of artificial neurons (otherwise called "nodes"). These nodes are associated with each other, and the quality of their associations with each other is doled out an esteem in view of their quality: restraint (most extreme being −1.0) or excitation (greatest being +1.0). In the event that the estimation of the association is high, at that point it shows that there is a solid association. Inside every node's constitution, an exchange function is implicit. There are three sorts of neurons in an ANN, **input nodes, hidden nodes**, and **output nodes** (Fig. 8).

incoming associations made up of the outgoing associations of different neurons. Therefore, the ANN should interface the yields of sigmoid units to the contributions of other sigmoid units.

8.2   One Sigmoid Unit

The diagram below shows a sigmoid unit with three inputs $\overrightarrow{x} = (x_1, x_2, x_3)$, one output, bias, and weight vector. Each of the inputs can be the output of another sigmoid unit (though it could also be raw input, analogous to unprocessed sense data in the brain, such as sound), and the unit's output can be the input to other sigmoid units (though it could also be a final output, analogous to an action associated neuron in the brain, such as one that bends your left elbow). Notice that each component of the weight vector corresponds to each component of the input vector. Thus, the summation of the product of the individual pairs is equivalent to the dot product, as discussed in the previous sections (Fig. 9).

8.3   Neural Networks Differ from Conventional Computing

To better comprehend artificial neural computing it is vital to know first how a traditional "serial" computing and its software procedure. A serial PC has a central processor that can address an array of memory areas where information and directions are stored. Calculations are made by the processor perusing a guideline and also any information the direction requires from memory addresses, the direction is then executed and the outcomes are stored in a predetermined memory area as required. In a serial framework (and a standard parallel one also) the computational strides are

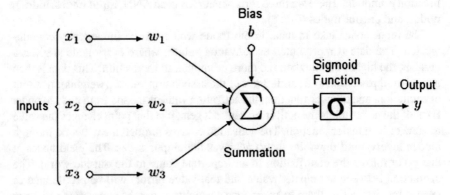

Fig. 9 Sigmoidal unit

deterministic, sequential and logical, and the condition of a given variable can be followed starting with one operation then onto the next. In examination, ANNs are not successive or essentially deterministic. There are no composite central processors; rather there are numerous straightforward ones which for the most part do simply take the weighted total of their contributions from different processors. ANNs do not execute programmed directions; they react in parallel (either recreated or real) to the pattern of information sources displayed to it. There are additionally no different memory addresses for putting away information. Rather, data is contained in the general activation "state" of the network. "Knowledge" is accordingly spoken to by the system itself, which is actually more than the entirety of its individual parts.

8.4 Applications of Neural Networks

Neural networks are all inclusive approximates, and they work best if the framework you are utilizing them to show has a high resistance to error. In any case they work extremely well for

- Capturing connections or finding regularities inside an arrangement of pattern;
- Where the volume, number of factors or decent variety of the information is exceptionally extraordinary;
- The connections between variables are enigmatically comprehended; or,
- The connections are hard to depict sufficiently with traditional methodologies [14, 15].

9 K-Nearest Neighbor (KNN)

This is a non-parametric procedure that can be used for both regression and order purposes. As we will probably gauge travel time (ceaseless ward variable) the focus will be essentially on the kNN regression. The kNN regression is a framework that in light of the k closest getting ready cases in the element space gives the yield forecast as the property estimation for the protest (normal of the estimations of its k closest neighbors). The part space is made in light of the autonomous factors which can be either nonstop or downright. More point by point depiction of the basic guidelines of the kNN regression can be found in [8] (Fig. 10).

One of the standard difficulties while using kNN method is the choice of k (neighborhood measure) as it can unequivocally affect the nature of gauge. For any given issue, a little estimation of k will incite a considerable distinction in desires and a broad regard may provoke a tremendous model slant. Writing proposes no right responses for finding the perfect size of k, however, rather to use the heuristic approach. Hence the cross-approval strategy was used. Cross-approval segments the information test into different v folds (self-assertively drawn, mixed up sub-tests

Fig. 10 K-Nearest neighbor

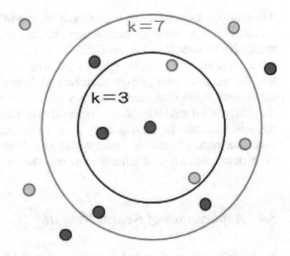

or parts). By then, for a settled estimation of k, the kNN technique is associated with make the conjecture on the v-th section (others are used as illustrations) and to assess the blunder [16]. This methodology is then progressively associated with each and every possible choice of v and diverse k. The esteem finishing the most lessened mistake is then picked as the incentive for k. In travel time estimating the principal educational accumulation was isolated into quarters, three of which were used for learning and one for testing. The separation between neighbors was processed in perspective of the squared Euclidian separation as portrayed by beneath condition where p and q are the request point and a case from the illustrations tests, independently.

$$D(p, q) = (p - q)^2$$

The 10-fold cross approval was utilized to choose the value of k for every street class independently. The search range for k was from 1 to 50, with the increment of one.

9.1 Applications of K-Nearest Neighbor

a. Travel time forecasting—Intertwined information from three datasets—Global Positioning System vehicle tracks, street systems foundation and meteorological information are utilized to assess the determining comes about. They are analyzed in the middle of various street classes with regards to supreme esteems, measured in minutes, and the mean squared rate blunder. The kNN regression method is utilized here. The estimation of neighborhood measure and arbitrarily drawn disconnected sub-test or portions are picked in an iterative way to arrange the street.

b. Healthcare disease prediction—For the most part, we respect the risk prediction demonstrate for cerebral dead tissue as the managed learning strategies for machine learning, which includes structured and unstructured data like age, gender, smoking status, personal information as input and output value gives whether the patient is suffering from cerebral infarction or not. The level of infection can also be known here. The k-nearest neighbor algorithm along with some other algorithm is used to for feature extraction in this situation.

c. Big data and machine learning for business management—Big data generated from different sources like social network, multiplayer's games, simulated worlds, mobile apps or different sensor data, if used in conjunction with machine learning algorithms can revolutionize the business industry in the following years if properly adopted by firms. K-nearest neighbor adapted for Map Reduce is a useful approach for classifying unstructured documents that are stored in file using a vector space model.

10　Dimensionality Reduction Algorithms

In machine learning and statistics, dimensionality decrease or dimension decrease is the way toward diminishing the quantity of random factors under thought by means of acquiring an arrangement of principal variables. It can be separated into feature selection and feature extraction [17].

10.1　Advantages of Dimension Reduction

We should take a gander at the advantages of applying Dimension Reduction process

- It helps in information packing and lessening the storage room required. It affixes the time required for performing same calculation/evaluation. Less dimensions prompts less calculation, additionally less dimensions can permit use of algorithms unfit for a large number of dimensions.
- It deals with multi-co-linearity that enhances the model execution. It expels redundant elements. For instance: there is no reason for putting away an incentive in two distinct units (meters and inches). It is useful in commotion expulsion additionally and as consequence of that we can enhance the execution of models.

11   Gradient Boost and Adaboost

Gradient boosting is a machine learning technique for regression and classification issues, which creates a forecast show as a gathering of powerless forecast models, ordinarily decision trees. It fabricates the model in a phase savvy form like other boosting strategies do, and it generalizes them up by permitting enhancement of a discretionary differentiable misfortune function. The possibility of gradient boosting started in the perception that boosting can be deciphered as an optimization algorithm on an appropriate cost function. Explicit regression gradient boosting algorithms were therefore created [18].

11.1   How Gradient Boosting Works?

Gradient boosting includes three elements:

A loss function to be optimized: The loss function utilized relies upon the kind of issue being solved. It must be differentiable, however, numerous standard misfortune capacities are upheld and you can characterize your own. For instance, regression may utilize a squared blunder and order may utilize logarithmic misfortune. An advantage of the gradient boosting model is that another boosting algorithm does not need to be inferred for every misfortune function that might need to be utilized, rather, it is a non-sufficiently specific system that any differentiable misfortune function can be utilized.

A weak learner to make predictions: Decision trees are utilized as the frail learner in gradient boosting. Specifically regression trees are utilized that yield genuine esteems for parts and whose yield can be included, permitting consequent models yields to be included and "remedy" the residuals in the forecasts. It is regular to oblige the feeble learner in particular courses, for example, a most extreme number of layers, nodes, parts or leaf nodes. This is to guarantee that the students stay feeble, however, can in any case be developed in a voracious way.

An additive model to add weak learners to minimize the loss function: Trees are included each one in turn, and existing trees in the model are not changed. A gradient descent technique is utilized to limit the misfortune while including trees. Generally, gradient descent is utilized to limit an arrangement of parameters, for example, the coefficients in a regression equation or weights in a neural system. In the wake of figuring blunder or misfortune, the weights are refreshed to lower that mistake. Rather than parameters, we have weak learner sub-models or all the more particularly decision trees. Subsequent to figuring the misfortune, to play out the gradient descent method, we should add a tree to the model that lessens the misfortune (i.e., take after the gradient). We do this by parameter zing the tree, at that point alter the parameters of the tree and move the correct way by decreasing the remaining misfortune.

11.2 Enhancements to Basic Gradient Boosting

Gradient boosting is a voracious algorithm and can over fit a training dataset quickly. It can profit by regularization techniques that punish different parts of the algorithm and for the most part enhance the execution of the algorithm by decreasing over fitting. In this segment we will take a gander at four improvements to basic gradient boosting

1. Tree Constraints
2. Shrinkage
3. Random sampling
4. Penalized Learning.

12 Conclusions

In this chapter, we have concentrated on the most recent progress over researchers with respect to machine learning for big data processing and techniques in the context of modern computing environments for various societal applications. Specifically, our aim to investigate opportunities and challenges of ML on big data and how it will affect the society. The chapter covers the discussion on ML in Big Data in specific societal areas.

References

1. Mohammad, N.R., Amir, E., Junhui, Z.: Machine learning with big data an efficient electricity generation forecasting system. Big Data Res. **5**, 9–15 (2016)
2. Hsieh, H.N., Chen, J.F., Do, Q.H.: A creative research based on DANP and TRIZ for an innovative cover shape design of machine tools. J. Eng. Des. **28**(2), 77–99 (2017)
3. Baldominos, A. et al.: A scalable machine learning online service for big data real-time analysis. In: IEEE Symposium on Computational Intelligence in Big Data (CIBD), pp. 1–8 (2014)
4. Liu, B., et al.: Scalable sentiment classification for big data analysis using Naive Bayes Classifier. In: IEEE International Conference on Big Data (2013)
5. Vaniš, M., Krzysztof, U.: Employing Bayesian networks and conditional probability functions for determining dependences in road traffic accidents data. In: Smart City Symposium Prague (SCSP) (2017)
6. Wang, Y.: Online machine learning for big data analytics by cognitive robots. In: Proceedings of the IEEE International Conference on Online Analysis and Computing Science (ICOACS 2016), Chongqing, China. IEEE CS Press (2016)
7. Chen, M.: Disease prediction by machine learning over big data from healthcare communities. IEEE Access (2017)
8. Šemanjski, I.: Analysed potential of big data and supervised machine learning techniques in effectively forecasting travel times from fused data. PROMET-Traffic Transp. **27**(6), 515–528 (2015)
9. Lee, C.K.M., Cao, Y., Ng, K.H.: Big data analytics for predictive maintenance strategies. In: Supply Chain Management in the Big Data Era, p. 50 (2016)

10. Pulipaka, S., Rajneesh, K.: Comparison of SOM and conventional neural network data division for PV reliability power prediction. In: 2017 IEEE International Conference on Environment and Electrical Engineering and 2017 IEEE Industrial and Commercial Power Systems Europe (EEEIC/I&CPS Europe). IEEE (2017)
11. Ali, M.U., Shahzad, A., Javed, F.: Harnessing the potential of machine learning for bioinformatics using big data tools. Int. J. Comput. Sci. Inform. Secur. 14(10), 668 (2016)
12. Vinay, A., et al.: Cloud based big data analytics framework for face recognition in social networks using machine learning. Procedia Comput. Sci. 50, 623–630 (2015)
13. Huijse, P., et al.: Computational intelligence challenges and applications on large-scale astronomical time series databases. IEEE Comput. Intell. Mag. 9(3), 27–39 (2014)
14. Lee, C.-R., Chang, Y.-F.: Enhancing accuracy and performance of collaborative filtering algorithm by stochastic SVD and its MapReduce implementation. In: IEEE 27th International on Parallel and Distributed Processing Symposium Workshops and Ph.D. Forum (IPDPSW). IEEE (2013)
15. Peters, D.P.C., et al.: Harnessing the power of big data: infusing the scientific method with machine learning to transform ecology. Ecosphere 5(6), 1–15 (2014)
16. Vallmuur, K., et al.: Harnessing information from injury narratives in the 'big data' era: understanding and applying machine learning for injury surveillance. Inj. Prev. 22(Suppl 1), i34–i42 (2016)
17. Zhang, W., Goh, A.T.: Multivariate adaptive regression splines and neural network models for prediction of pile drivability. Geosci. Front. 7(1), 45–52 (2016)
18. Lin, J., Alek, K.: Large-scale machine learning at twitter. In: Proceedings of the 2012 ACM SIGMOD International Conference on Management of Data. ACM (2012)
19. Zhou, Q., Shao, X., Jiang, P., Gao, Z., Zhou, H., Shu, L.: An active learning variable-fidelity metamodelling approach based on ensemble of metamodels and objective-oriented sequential sampling. J. Eng. Des. 27(4–6), 205–231 (2016)
20. Kremer, J., et al.: Big universe, big data: machine learning and image analysis for astronomy. IEEE Intell. Syst. 32(2), 16–22 (2017)
21. Moran, M.S., et al.: Agroecosystem research with big data and a modified scientific method using machine learning concepts. Ecosphere 7(10), (2016)
22. Törönen, P., et al.: Analysis of gene expression data using self-organizing maps. FEBS Lett. 451(2), 142–146 (1999)
23. Ruta, D.: Automated trading with machine learning on Big Data. In: IEEE International Congress on Big Data (BigData Congress) (2014)
24. Yoo, C., Luis R., Juan L.: Big data analysis using modern statistical and machine learning methods in medicine. Int. Neurourol. J. 18(2) (2014)
25. Haupt, S.E., Kosovic, B.: Big data and machine learning for applied weather forecasts: forecasting solar power for utility operations. In: IEEE Symposium Series on Computational Intelligence (2015)
26. Diaconita, V., Alexandra Maria, I., Iuliana, D.: Big data and machine learning for business management
27. Diaconita, V.: Big data and machine learning for knowledge management. In: Proceedings of the 9th International Conference on Business Excellence (2014)
28. Wang, X., et al.: Building efficient probability transition matrix using machine learning from big data for personalized route prediction. Procedia Comput. Sci. 53, 284–291(2015)
29. Hammond, K., Aparna, S.V.: Cloud based predictive analytics: text classification, recommender systems and decision support. In: IEEE 13th International Conference on Data Mining Workshops (ICDMW). IEEE (2013)
30. Rahman, M.N., Amir, E., Junhui, Z.: Machine learning with big data an efficient electricity generation forecasting system. Big Data Res. 5, 9–15 (2016)
31. Iniesta, R., Stahl, D., McGuffin, P.: Machine learning, statistical learning and the future of biological research in psychiatry. Psychol. Med. 46(12), 2455–2465 (2016)
32. Semanjski, I.: Potential of big data in forecasting travel times. Promet-Traffic Transp. 27(6), 515–528 (2015)

33. Van Quan, N., et al.: Real-time earthquake detection using convolutional neural network and social data. In: IEEE Third International Conference on Multimedia Big Data (BigMM). IEEE (2017)
34. Li, L., et al.: Risk adjustment of patient expenditures: a big data analytics approach. In: IEEE International Conference on Big Data. IEEE (2013)
35. De Vlaming, R., Groenen, P.J.F.: The current and future use of ridge regression for prediction in quantitative genetics. BioMed Res. Int. (2015)
36. Wen, L., et al.: The impacts of river regulation and water diversion on the hydrological drought characteristics in the Lower Murrumbidgee River, Australia. J. Hydrol. **405**(3), 382–391 (2011)
37. da Pimentel, E.C.G., et al.: Use of ridge regression for the prediction of early growth performance in crossbred calves. Gen. Mol. Biol. **30**(3) 536–544 (2007)
38. Schimbinschi, F., et al.: Traffic forecasting in complex urban networks: leveraging big data and machine learning. In: IEEE International Conference on Big Data (Big Data). IEEE (2015)
39. Armes, T., Refern, M.: Using Big Data and predictive machine learning in aerospace test environments. In: AUTOTESTCON, 2013. IEEE (2013)

Personalized Diabetes Analysis Using Correlation-Based Incremental Clustering Algorithm

Preeti Mulay and Kaustubh Shinde

Abstract Digested food is used as energy; the body breaks down carbohydrates from the food into glucose which is absorbed into the body using hormone insulin. This insulin is helpful in reducing the risk of diabetes. Diabetes Mellitus is a disorder of metabolism; it is one of the highest occurring diseases in the world, having affected over 422 million people. Diabetic level in person depends on various factors; if their values are kept in control, a diabetic patient can improve his/her expectancy of life. The trends in the diabetic levels of a person can be extracted using data mining techniques from individual person's medical reports. *Personalization* of these medical reports of an individual will produce apt analysis for diabetologist, considering medical history from inception, showing progress of treatment. Such customized approach is categorically beneficial for not only medical professionals' like the doctors, nurses, medical practitioners, etc. but also for an individual, their families, researchers and entire society at large. This chapter describes the details about incremental clustering approach, Correlation-Based Incremental Clustering Algorithm (CBICA) to create clusters by applying CBICA to the data of a diabetic patients and observing any relationship which indicates the reason behind the increase of the diabetic level over a specific period of time including frequent visits to healthcare facility. These obtained results from CBICA are compared with results obtained from another incremental clustering approaches, Closeness Factor Based Algorithm (CFBA), which is a probability-based incremental clustering algorithm. 'Cluster-first approach' is the distinctive concept implemented in both CFBA and CBICA algorithms. Both these algorithms are 'parameter-free', meaning only end-user requires to give input dataset to these algorithms, clustering is automatically performed using no addition dependencies from user including distance measures, assumption of centroids, no. of clusters to form, etc. This research introduces a new definition of outliers, ranking of clusters and ranking of principal components. Scalability: Such personalization

P. Mulay (✉) · K. Shinde
Department of CS, Symbiosis Institute of Technology, Symbiosis International (Deemed University), Pune, India
e-mail: preeti.mulay@sitpune.edu.in

K. Shinde
e-mail: kaustubh.shinde@sitpune.edu.in

© Springer Nature Singapore Pte Ltd. 2019
M. Mittal et al. (eds.), *Big Data Processing Using Spark in Cloud*,
Studies in Big Data 43, https://doi.org/10.1007/978-981-13-0550-4_8

approach can be further extended to cater the needs of Gestational, Juvenile, Type-1 and Type-2 diabetic prevention in society. Such research can be further made distributed in nature so as to consider diabetic patient's data from all across the world and wider analysis. Such analysis may vary or can be clustered based on seasonality, food intake, personal exercise regime, heredity and other related factors. Without such integrated tool, the diabetologist in hurry, while prescribing new details, may consider only latest report, without empirical details of an individual. Such situation is very common in these stressful and time-constraint life, which may affect the accurate predictive analysis required for the patient.

1 Introduction

In the era of automation or 'Software Product Line Management (SPLM)' 'Mass Customisation' is the popular term coined by researchers. 'Mass Customisation' focuses on customized products used by masses. Even though this concept sounds completely controversial few years ago, but now this is the trend of present years and future too. Due to huge amount of seasonality changes, behavioural aspects, food and exercise habits + regime, heredity, work-culture and society at large, diabetes became a major concern. This disease may start taking shape from Gestational days follows through Juvenile stages and then Type-1 and 2 etc. If the diabetes follows such life cycle phases, then it becomes mandatory to perform following iteratively:

- Medical check-ups,
- Pathology lab tests,
- Monitoring of diet + exercise,
- Food restriction,
- Continuous medications, etc.

Diabetologist may find it difficult without the support of IT healthcare tools, to cater the needs of every individual by spending quality time to surf through the empirical details, before prescribing medicines. This is the focus of the research. Personalization is the key, as same prescription given to two different diabetic individuals may prove harmful. To achieve said Personalization, it is necessary to cluster the diabetic patients' details and keep incrementally learning about the variations in them, due to multiple factors. Such incremental learning achieved via incremental clustering proves the unique solution for diabetologist of this generation. The number of factors to consider before prescribing medication and looking at availability of time, it is mandatory to propose correlation—[11] based incremental clustering solution for implementing Personalization for the diabetic prevention and cure. This research can then be extended to apply for any other kind of diseases mushrooming these days or may, in future.

Probability-based Incremental Clustering given by CFBA [7, 8] gives relationship/mapping details only between 0 and 1, as the probability range is defined in such way. But to increase the horizons and append negative compatibility details in analysis, it is mandatory to shift the research focus towards Correlation-based Incremental Clustering using diabetes data, fit individuals data, young grads data, etc.

In addition to achieving Personalization, this research redefined the concept ranking attributes too. Incremental clustering works in three phases, as follows of Outliers, ranking clusters and

- Basic formation of clusters phase,
- Update existing cluster on influx of data,
- Form new cluster on arrival of new data.

On arrival of new data, when the incremental clustering algorithm decides to either update existing cluster or form a new cluster, the cluster ranking may be affected. Clusters are ranked in this research by frequency of usage of cluster. The cluster which is given a zero or negative ranking, may indicate never-used cluster, after formation of it. Such clusters may contain cluster members/data series which are not useful for the analysis of results. Such non-useful data series cluster can be considered as having extremely low or high/rarely used values and hence renamed as Outliers.

In this pre-clustering phase, data preprocessing takes place, by using any of the following methods, based on requirement, type of raw data, size of data-attributes and goal of research:

- Principal Components Analysis (PCA),
- Independent Components Analysis (ICA),
- Practical Swarm Optimisation (PSO),
- Feature Selection (FS),
- Feature Extraction (FE),
- Genetic Algorithms (GAs), etc.

PCA's duty is to give/select or suggest Principal Components only. Basic clusters are formed based on highly varied attribute. Such highly varied attributes are researched during all iterations to form highly ranked attribute over various iterations of clustering. Any preprocessing techniques used to be applied only once to extract highly varying attributes, so as to achieve quality clusters. But while working on Personalisation research work using diabetes data, it is observed and implemented that ranking of clusters and attributes at every iteration became a mandatory part of incremental learning via incremental clustering.

At present Personalisation of diabetes patients is achieved using standalone systems. A system which executes CBICA executes influx of data to

- form clusters,
- append clusters,
- edit clusters,
- produce analysis,
- form outliers,
- assign ranking of clusters and attributes,

- map classes to clusters,
- Naming clusters, etc.

The next phase of this research includes formation of distributed CBICA (D-CBICA). This D-CBICA will work in two phases, namely:

- D-CBICA forming clusters centrally, as the algorithm is available with central server,
- D-CBICA forming clusters globally, as the algorithm is available with every node of distributed system.

Another thought about extension includes the weight-assignment strategy design, to effectually handle the non-numeric data on arrival. In this case, again the PCA or alike algorithm will be executed to know at initial phase, the Principal Components/Impactful attributes. If, PCA results in selecting a non-numeric attributes and such attributes will prove the most important contributor while forming, appending or editing clusters, then it is necessary to have weight-assignment matrix ready to assign numeric equivalent values for further processing of data series in dataset.

CBICA is a parameter-free incremental clustering algorithm proved very useful for Personalisation of diabetic patients already. D-CBICA will be able to correlate world-wide data details to further prevent this wide-spread disease. Rather Personalisation proves very essential to detect diabetes or any other disease in very early stages due to its varied features.

Learning about diabetic patients using incremental-clustering concepts is achieved in this research. Incremental clustering algorithms are applied on iteratively generated diabetic patients' data to learn about variations in glucose levels based on impactful attributes. This research is carried out in two phases. In the first phase the probability-based incremental clustering algorithm is applied on diabetic patient data to understand how clustering is applied, how incremental clustering is achieved and driven by which impactful attributes in various iterations. In the second phase of research the same dataset of diabetic patients is executed with correlation-based incremental clustering algorithm. Correlation matrix is used as it shows the negatively correlated attributes and in turn the appropriate clustering which is missing in probability-based incremental clustering approach. To conclude these two approaches are compared to understand the apt way to learn about appropriate machine learning technique for diabetic data analysis. The order-independence is implemented to verify the incremental nature of clustering algorithm. The cluster database is created and maintained to accommodate influx of new data, append existing clusters with new data series and to create new clusters as and when required. To perform the learning based on diabetic patients, various clinical datasets are used for this study showing the detail of patients. 'Closeness Factor Based Algorithm (CFBA)' probability-based error-based statistical incremental clustering algorithm is applied at first phase of study. Correlation matrix is replaced with probability in the modified CFBA to achieve new sets of comparative results. The key findings include learning about impactful attributes of diabetic patient data which is used for prevention of this disease to achieve wellness of individuals, is the focus of this research study. Precision, recall, f-measure are applied to other incremental clustering algorithms also like Cobweb and I k-means is used for validation of results. The decision to whether update existing cluster or form a new cluster on arrival of new data is based on representative of cluster, threshold range of cluster and also to weights assignment matrix used for non-numeric attributes of input dataset.

To further increase the capabilities of D-CBICA, it is necessary to focus on privacy and security of individual's data, medical reports, etc. 'Ants feeding Birds' algorithm will play a major role in it. 'Ants feeding Birds' will enhance the distributed capabilities of D-CBICA by automatically fetching the patients' documents from all sorts of connected devices whether thin or thick, Internet-based or dew-computing-based or IoT-based devices, continuously. The secured handshaking capability inherent in 'Ants feeding Birds' algorithm will allow the distributed algorithm to fetch required document using either cloud based/internet-based systems or dew-computing based/mobile network-based systems, for analysis of data. As this process of fetching medical reports is done automatically and frequently, the diabetologist will have analysis of updated details of patients before the actual clinic visit by patient. This will produce more 'feel good' hormones not only in patient to help him cure fast, handle disease easily but also to the busy doctor to handle many patients effactually in given restricted time frame.

These days IoT based, 'Cloud of Things', 'Internet of Everything (IoE)', etc. produces a huge amount of time series data [2, 6]. It is necessary to have distributed systems which recommends apt incremental clustering algorithm and best distance measures to end-user automatically, instead of 'trial and error' methods. Sneha et al. (2017) published the work of recommendations effactually, to incorporate more professional and pioneered incremental clustering algorithms which are not parameter-free like CFBA, CBICA and D-CBICA. Inclusion of such algorithms improves validation of results and show comparative analysis.

It is also feasible that area-wise, state-wise, country-wise, age-wise, gender-wise, daily-activity-wise the results may differ. Different incremental clustering algorithms may prove suitable based on above-mentioned parameter. To finally conclude, Personalisation is achieved at all levels of clustering, at preprocessing level PCA given attributes are confirmed at post-clustering phase, by analysing the formed clusters based on which specific attribute at post-clustering phase, the diabetologist can prescribe medication by either combining two or more clusters or dividing one large cluster into multiple pieces. Negatively correlated members in the cluster should be the major focus while treating the patient. These members may prove important for an individual who is diabetic or not.

A true distributed system of its kind is proposed based on machine learning concepts in homeopathy too. This distributed system is an integrated concept of incremental clustering and bio-markers. These bio-markers are the systems available with homeopath in his/her mobile device/iPad/any thin client. these bio-markers are capable of finding varied hidden diseases including stress, heart diseases, mood swings, etc. such analysis given by bio-markers combined with D-CBICA will prove a unique distributed system. Sentilizer is the proposed name of the system. This system will work on empirical data of patient starting from history taken by homeopath during first visit. Iterative details of patients, falling-ill pattern, category of illness, seasonality changes, behavioural changes, circumstances, etc. is noted by this system. This system again varies from Cloud Computing environment to Dew-Computing environment, by considering Internet-based as well as mobile network-based systems. Fog-Computing environment will handle the security and privacy of data while handling iteratively. Incremental learning using Sentilizer can be achieved and useful for not only the homeopath, but also to patient, family, medical and engineering interns, researcher from both fields/domains, society at large.

Homeopathy medicines are inherently classified as journal medicines, chronic medicines, bio-medical options and flower-remedies, etc. there could be many other classifications feasible doctor-wise. Homeopathy by nature cures patients at physical as well as emotional levels to achieve overall wellness to provide help support while handling increasing number of patients per doctor, it is mandatory to have such automatic healthcare–IT blended system. Analysis given by Sentilizer will be available for doctor for reference at any time, being supported by all types of computing. This will help the doctor to pull the required data securely as and when needed, might be in case of emergency or after clinic hours, with complete proof of prescribed details till date, handing over the patient to other surgeon or anyone else becomes easy.

Such distributed system can be extended further with maths-based systems like CueMaths, wherein students may relocate from places to places, school to school, teacher to teacher and area to area (including varied countries due to parents' job these days).

Smart Supply Chain Management in Homeopathy also proved a very useful distributed system indeed, especially based on faith of the homeopath.

2   Related Work

While working on couple of integrated research papers on Homeopathy and Machine Learning, I sincerely thought that all upcoming Homeopathy students need to learn Information-Technology related subjects too. Technology provides a great support to healthcare experts to better analyse patient in hand, in this 'data driven life'. Technology provides reach to individual empirical data with ease and hence prescribing medicines based on factual numbers along with subjective and intuitive sensations within. Bio-markers these days captures all possible hidden truths and symptoms very easily which can be added to experience of healthcare professional more effectually. In fact all healthcare professional courses now should include technology based courses and psychology-based optional courses too. To personalize integrated system for diabetologist, there should be amalgamation of numbers in terms of pathology tests over the period of time, inputs from bio-markers and also optional place for healthcare professional to note the psychological details too. So this proposed integrated system will have following modules:

- Text mining

- Sentiment analysis
- Opinion mining
- Empirical data (numbers given by pathology tests)
- Empirical data about treatments given by diabetologist
- Daily routine of individual patients (food + exercise)
- Bio-markers
- Data clustering techniques (to form clusters for self-learning, apt diagnosis).

Sharon (2017) [12, 18] In this paper, the author talked about personalized health-care systems at large. The author primarily focused on use of 'self-tracking of health' by individuals. Author also elaborated the various concepts like Sensing-technology, self-narratives, digital-storytelling, digital-identity, 'self-knowledge through numbers', 'the data driven life', 'health data is a new natural resource'. The author discussed about the varied formats of all these tracking systems and amalgamation of data may be difficult if required to do so. Yes, this will be the major concern if diabetologist will believe on or use the data collected by such self-tracking system for personalized prescriptions designs. Instead it will be better to start self by healthcare professional and collect, store, sort, cluster, use, re-use all possible patients/individual's data before prescribing medicines with every visit of an individual patient.

Ada et al. (2017) in their detailed medical write-up discussed about '*Overtreatment in Diabetes Management*' with more focus on comparative effectiveness of personalized [4, 13, 14] medicine for a variety of chronic diseases. Dr. McCoy said in the write-up that there's a substantial fraction of patients with diabetes who are potentially over treated. Dr. McCoy discuss about following details, 'what constitutes overtreatment, how overtreatment can harm patients and increase their disease burden, and examine some of the reasons for overtreatment. Many of the studies she'll discuss show that intensive treatment and potential overtreatment are common in routine clinical practice, disproportionately affecting older and frailer patients, and directly lead to increased risk of hypoglycaemia.' She further stated other effects of personalized and overtreatment in diabetes handled alone by medical professional without help of technology. To provide helping hand to medical practitioners and specialists having variety of patients with huge numbers, integrated personalized learning system is a must.

eGlycemic Management System®, or eGMS®, 2017, is one of the leading patented systems by Glytec. This system proved the successful improvements in normal to acute diabetes. Glytec's eGMS® facilitates the personalized therapy management for diabetes patients. This system effectually reduces various factors affecting diabetes including hypoglycaemia, haemoglobin A1C, etc. The focus of this system is to primarily measure and control diabetes.

Glytec is one of the leaders in pioneering personalized diabetes management software. This software by Glytec is patented and FDA approved. This software is useful for both inpatient and outpatients with the aim of achieving or preventing glycaemic levels all across the globe. Glytec also have a specialized module for glycaemic management and decision support system which is cloud-based called

Fig. 1 a and **b** Shows personalized system developed by Glytec, Glytec Cloud based system *Source* http://www.glytecsystems.com/Solutions.html

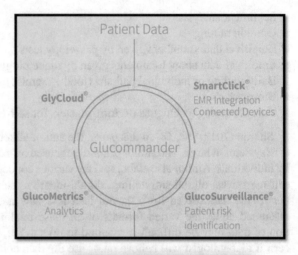

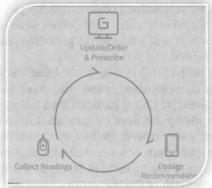

eGMS®. This cloud-based platform, as shown in Fig. 1, provided support to all involved entities including patients, researchers, providers etc. to monitor and control rather reduce the frequency, risk and cost of both hyperglycaemia and hypoglycaemia. As the name suggest Glytec, it proved a true integrated service provider for the society at large for preventing Glycaemic levels with the help of technology. Glytec quotes on their website '*eGMS® interventional clinical decision support tools (including Glucommander™ suite of FDA-cleared proprietary algorithms for IV, subcutaneous and paediatric insulin dosing, SmartClick®, robust analytics by GlucoMetrics® and surveillance using GlucoSurveillance® capabilities etc.), allow providers to standardize processes and personalize treatment across all areas of care, and are proven to result in significant improvements in clinical and financial outcomes over traditional methods*'. Glytec's system is compatible with all types of clients, thin and thick, as shown in the Fig. 2.

Bertsimas et al. [3] The authors have discussed in their research about managing type 2 diabetes effectually. They claim that patient specific factors do not affect man-

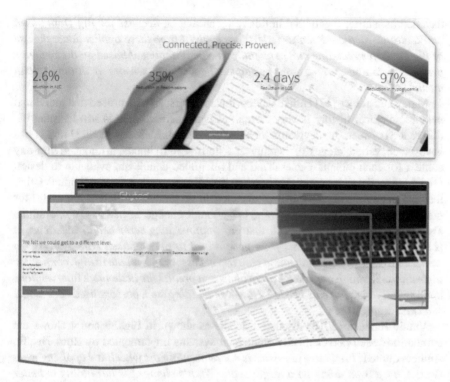

Fig. 2 **a** and **b** Shows the Glytec system's web screen shot, to show integration of all possible clients, thick and thin *Source* http://www.glytecsystems.com/

aging type 2 diabetes. An extensive data driven algorithm is customized and made to achieve personalized diabetes management to improvise health outcomes. They have derived their methodologies and algorithms based on empirical data from 1999 to 2014 for almost 11 k patients with type 2 diabetes. These methodologies and algorithms are based on 13 pharmacological therapies. Analysis of range of outcomes per patient is achieved using k-nearest neighbour algorithm. Authors quoted '*The neighbours were chosen to maximize similarity on individual patient characteristics and medical history that were most predictive of health outcomes. The recommendation algorithm prescribes the regimen with best predicted outcome if the expected improvement from switching regimens exceeds a threshold. We evaluated the effect of recommendations on matched patient outcomes from unseen data*'. to summarize, this personalized approach used by authors in their research study proved substantial improvements in HbA$_{1c}$ outcomes as compared to given standards. Further their research and recommendation system can be useful for various service providers to inform about diabetic care details.

Cheryl and Lidong (2017) used big data concept in healthcare due to growth of heterogeneous and unstructured data and difficulty in processing such massive data. Big data can be the answer for required processing and personalization of healthcare data details. author mentioned in this research that '*A new use for Big Data based on Electronic Medical Records (EMR) has evolved from these applications: the use of Big Data to streamline and personalize medical treatment based on the patient's previous data and current data, perhaps collected with sensors or through social media*'.

Malay and Teresa [17] Unlike Glytec, this set of researchers proved that individual persons' medical history is very crucial, as healthcare domain is still learning the difference between the keywords 'healthy' and 'normal'. Many medical text books have given varied ranges of healthy and normal individuals, without giving any context for each patient. Generalized and integrated details are available to design predictive models it is essential to know personalization concept per individual. Other than patient care and prevention of diabetes, these predictive models are useful for preventing readmission in hospitals, behavioural change in patients, etc. Authors also mentioned in their research that: *To improve long term health outcomes, it is necessary to shift the focus from generalised to personalised care systems. To achieve this kind of personalisation care it is necessary to implement high confidence algorithms. The high confidence algorithms can predict the actionable interventions and suggest improvements needed. However, building such personalised care model is a challenging task.*

Vinnie Ramesh (2016) discussed details as shown in Figs. 3 and 4 shows the correlation-based personalized commercial systems implemented by Rock Health. Authors quoted '*Predictive analytics is not reinventing the wheel. It's applying what doctors have been doing on a larger scale. What's changed is our ability to better measure, aggregate, and make sense of previously hard-to-obtain or non-existent behavioural, psychosocial, and biometric data. Combining these new datasets with the existing sciences of epidemiology and clinical medicine allows us to accelerate progress in understanding the relationships between external factors and human biology—ultimately resulting in enhanced reengineering of clinical pathways and truly personalized care.*" [22].

Authors in this (Philips et al. 2013) research paper mentioned about economic evaluations of personalized medicines in general, without focusing on a specific disease. They have carried out extensive research at the University of California, San Francisco Centre for Translational and Policy Research on Personalized Medicine (TRANSPERS). They have discussed about following major points in this research and they are

- Challenges involved in implementing personalized medications,
- Different evaluation approaches required for personalized medications,
- Key factors in personalized medicine,
- Consideration of patients' heterogeneity,
- Real-world perspective and diverse populations while analysing economics details [18],
- Usage of emerging technologies for personalized medicines,
- Integration of behavioural economics into value assessments and
- Application of various methods for assessing economics details.

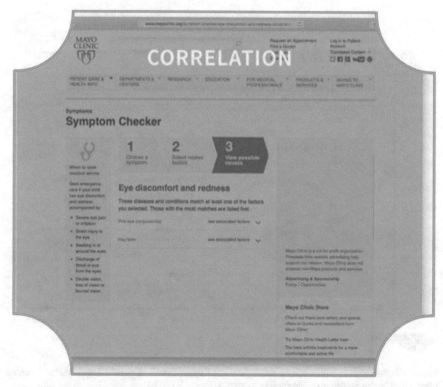

Fig. 3 Shows Rock Health's screen shots showing useful correlation for implementing personalized system *Source* https://rockhealth.com/reports/predictive-analytics/

Shobhita (2017) said that service providers of personalized medicine cannot work effectually without the support and willingness of patients, their feedback and incremental learning about delivery, experience and demand. Providers will be able to standardize their offering in continuous fashion based on all the above, given by patients and users of services. Consumer-driven healthcare to continue it is necessary that patient share their personalized healthcare experience and demand more for it. They quoted in this paper '*To address the demand for personalization, providers segment their patient populations. As suppliers and service providers, you need to show that not only do you understand your provider customers' diverse patient segments, but that your offerings can directly support these different groups. This way, providers will be able to think of you as partners in their efforts to achieve care personalization, rather than barriers to doing so*' [19].

According to these set of authors and researchers, with the help of standardized clinical processes related to multiple diseases, and blooming technology, healthcare experts and physicians will be able to give more valuable quality time to each patient. This will not only increase the faith on that physician by patient but physician will be able to read more details about individuals. This will generate personalized sys-

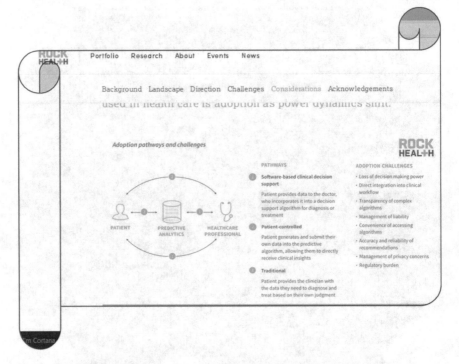

Fig. 4 Shows Rock Health's screen shots showing further details of personalized system *Source* https://rockhealth.com/reports/predictive-analytics/

tems with better preventive measure without any doubts. Multiple objectives can be achieved with the help of personalized technology, patients receive apt treatments at right time with extra care. Physicians can think more creatively and design personalized treatment plan, by keeping an eye not only on physical ailments and issues, but also other personal details including behaviour, eating pattern, exercise regime, social aspects, sleeping styles, etc.

Other than medical field personalization using technology is used enormously. One of the fields of such study includes schools and related activities. To be frank, these personalization practices with or without technology should start from inception of individual itself, not when he/she becomes patient. It is kind of mandatory to have real empirical data about behaviour, social aspects and other details to predict and perform analysis better. John et al. [20] as mentioned in one of their research publications, that, several lessons are learned after implementing personalized learning in school and they are

- Learning needs of each individual student is captured to provide required support in flexible way,
- Without affecting traditional school practices, personalized learning approaches were implemented with varying degrees,

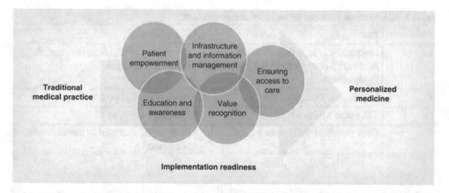

Fig. 5 Shows screen shot by future medicine showing progressive tasks *Source* https://www.futuremedicine.com/doi/full/10.2217/pme-2016-0064 (Daryl et al. 2016)

- Authors quoted 'There is suggestive evidence that greater implementation of PL practices may be related to more-positive effects on achievement'.

Daryl et al. (2016) said in their research paper that '*Research and innovation in personalized medicine are surging, however, its adoption into clinical practice is comparatively slow. We identify common challenges to the clinical adoption of personalized medicine and provide strategies for addressing these challenges. We categorized challenges into five areas of need: education and awareness; patient empowerment; value recognition; infrastructure and information management; and ensuring access to care. We then developed strategies to address these challenges. In order for healthcare to transition into personalized medicine, it is necessary for stakeholders to build momentum by implementing a progression of strategies.*' Figure 5 shows the life cycle of implemented system by authors.

Researchers have proposed various incremental clustering algorithms over the period of time, but these algorithms have some limitations. To overcome these, the proposed system clusters the various data sets available.

Clustering of various types of data sets is not achieved by any incremental clustering algorithm. There are various algorithms available which do clustering for specific data set. Multi-Assignment Clustering (MAC), for clustering Boolean data that can concurrently be part of several clusters [5]. An expectation-maximization (EM) algorithm is presented where the source framework of the clusters and the cluster memberships of each data item are concurrently estimated (Frank et al. 2012). CFBA attained all the attributes necessary for an almost 'parameter-free' algorithm [7–9]. Scaling of dynamically available data is comfortably achieved using CFBA. CFBA converges comfortably being 'cluster-first' approach and not 'centre-first'.

Incremental clustering algorithm rooted on closeness, a systematic and scalable approach which updates clusters and learns new information effectively (Mulay and Kulkarni 2013). The alpha-numeric data, which was used, was first transformed into numeric data, using the weight-assignment algorithm.

Table 1 Comparison between probability and correlation

Sr. No.	Probability	Correlation
1	Probability here is the ration between each series (sample) and the grand total (population)	Correlation will give us the relation between two series
2	It is an unbiased approach	It is a biassed approach
3	The range is 0 to +1	The range is −1 to +1
4	Only positive relation can be obtained	Both positive as well as negative relation is obtained

A two-level fuzzy clustering approach to obtain the dimensionality reduction [1]. Here the prototypes are referred as the basic clusters which are generated initially, and then further used to update or append the clusters after influx of new data.

A comparative study about shaping cluster structure and is based on selecting an optimal threshold value to form a cluster (Mulay 2016). Incremental clustering approach 'Incremental clustering using Naïve Bays and Closeness-Factor' (ICN-BCF).

A new Density-Based Clustering (DBC) algorithm was proposed considering correlation coefficient as similarity measure [10]. This algorithm though computationally not efficient, is found to be effective when there is high similarity between patterns of dataset. The computations associated with DBC based on correlation algorithms are reduced with new cluster merging criteria.

3 Mapping of Probability and Correlation in Incremental Clustering

Considering the work done related to the incremental clustering algorithms, the possible enhancement is to replace the probability-based similarity measure with the non-probability-based similarity measures. The most appropriate non-probability-based similarity measure is the Pearson's coefficient of correlation.

Table 1 gives us the idea, why probability can be replaced by correlation. The range of correlation gives the positive as well as the negative relation between the data series, which is an added advantage over the probability which only gives us the positive relation between the data series. The correlation has a biassed approach, which gives equal importance to all the data series in the input data.

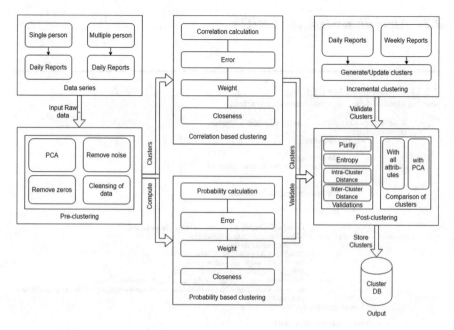

Fig. 6 Shows architecture of personalized CBICA

3.1 Architecture of Personalized CBICA

The CBICA is designed in such a way that, it can be used to incrementally cluster available data of various people.

In Fig. 6, initially a dataset, amongst the various available datasets, is taken as a raw data, as an input to the system. This input raw data is converted into the numeric data, using the weight-assignment algorithm. Further the preprocessing of data is done, by any of the preprocessing techniques available. In the next step the CBICA is applied to the preprocessed data. As this algorithm uses a 'cluster-first' approach, user is not responsible for defining the number of clusters. The basic clusters are generated initially. These clusters are saved into the cluster database. Due to the 'cluster-first' approach, hoping of cluster members is not possible. If there is an influx of new data, the incremental approach can be used. The influx of new data will consists of daily or weekly reports. The incremental approach uses all the above steps to update the existing clusters or generates new clusters if needed. These updated or newly generated clusters are also saved into the cluster database.

The next important step is to validate the clusters, in the post-clustering phase. Statistical validation techniques are used to validate the numeric data; whereas conceptual validation techniques are used to validate the categorical data. Various validation techniques are used to validate the clusters such as Purity, Entropy, Intra-Cluster Distance and Inter-Cluster Distance.

3.2 Algorithm Steps for CBICA

3.2 Algorithm steps for CBICA:
Knowing what the future holds is very important in healthcare.

- Input: Read the .csv input file – store the instances in a ArrayList.
- If create new cluster then
  - Method = newClusters()

    - Correlation()
      For all series, calculate the correlation value

    - calCloseness()
      For all series, compute the closeness value

    - createCluster()

      - Create clusters using the closeness and correlation values,
      - Maintain cluster details in cluster-db,
      - Calculate threshold value,
        Threshold = Average of all closeness values.
      - For all series, set processedFlags [] = 0.
      - If flag ≠ 0, series is already part of a cluster.
      - Else create a new cluster and add this series as part of the new cluster.
      - At the end of for-loop, each series should be part of some cluster, and all the series should have processedFlags ≠ 0.
- If clusters are formed Incrementally, then

  - Method = incrementalClusters()

    - Correlation()
      Calculate correlation values for all series from the influx of new data using step 2.a.i.
    - calCloseness()
      Calculate closeness values for all series from the influx of new data using step 2.a.ii.
    - updateCluster()
      For each existing clusters,
      - Get each series in existing cluster, S(i).
      - Get the closeness value of S(i).
      - For each newly added series,
        if processedFlags ≠ 0, ignore series.
      - At the end of the for-loop, each series should be part of some cluster, and all the series should have processedFlags ≠ 0.
        If threshold values does not match then form a new cluster.
- Output: Write output in the given .csv file.
  This file will contain the attributes, cluster number and the closeness value of each series.
- End.

3.3 Results Obtained by Implementing Personalized System

The dataset used here is a pathological report of Diabetes Mellitus, which is a numeric dataset, containing of eight attributes and 15,000 instances. The available dataset undergoes a data cleansing process in the pre-clustering stage, where various pre-clustering techniques, such as: Remove Zero, Remove Duplicate and Principal Component Analysis, are applied on the input dataset.

In the main clustering stage, the input data is given to the algorithm to obtain the clusters. The clusters are obtained after a series of calculations and computations.

CBICA was applied to the diabetes patient's datasets [15, 16] taking into account eight attributes. These attributes were chosen based on common occurrence in every

Fig. 7 Shows Analysis of reports for person P1 using correlation based algorithm

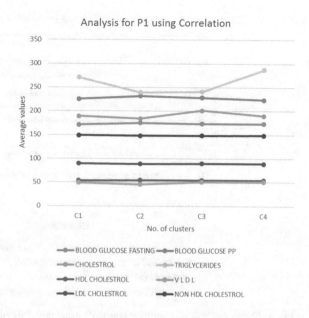

diabetic patient's report and their relative importance with respect to factors triggering the disease.

The study found out relations between various attributes and carried out outlier analysis wherein we identify which attributes are the underlying causes for a data series to be an outlier.

The idea of personalizing these reports of an individual person was born as day-by-day, doctors are getting busier, as the number of patients are increases all the time. Thus doctors do not completely check the history of the patients. This can cause improper treatment to the person.

To avoid such mishaps, this system will help doctors to get the analysis of the recent medical reports of the person. This analysis will save the doctor's time and will increase the efficiency as well.

Here the recent 100 reports of a person P1 were considered for doing the analysis. Each of the reports had eight attributes. Figure 7 is the analysis of results for P1 using CBICA. Four clusters were obtained using the correlation-based algorithm. Figure 7 represents the average values of each attribute for the corresponding clusters.

Figure 8 is the analysis of results for P1 using CFBA. Four clusters were obtained using the probability-based algorithm. Figure 4 represents the average values of each attribute for the corresponding clusters.

Both the figures clearly suggest that out of the eight attributes, Triglycerides is the most impactful attribute causing diabetes to the person. There is a continuous variation in the values of Triglycerides. Considering these results, doctor can easily medicate the patient for controlling the Triglycerides, without wasting much time.

Table 2 shows the application of correlation-based incremental clustering algorithm on diabetic patients' data set. This table show that in the first iteration of

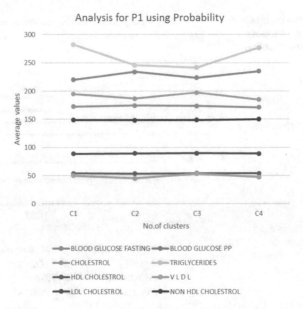

Fig. 8 Shows Analysis of reports for person P1 using probability-based algorithm

Table 2 Diabetic patients' dataset

Dataset size	No. of iterations	Threshold	No. of clusters	Cluster members
4898	1 (1–4898)	0.040817	3	C1: 2138, C2: 2758, C3: 2
4898	2 (1–2449)	0.037309	2	C1: 1148, C2: 1301
	(2450–4898)	0.043468	4	C1: 729, C2: 525, C3: 1193, C4: 2
4898	1 (4898–1)	0.040817	3	C1: 2238, C2: 2656, C3: 4

execution, the entire data series form 1 to 4898 were considered and three clusters were formed with threshold value and cluster member details as shown. To prove the incremental nature of this newly developed correlation-based clustering algorithm, two iterations are executed with different set of data series and the details including number of clusters, members, etc. are as shown in the Table 2. The order-independence property of incremental clustering algorithm is proved in the third iteration, as shown in the Table 2. The datasets were given as input in reverse order to confirm the correlation-based incremental clustering algorithm produces same quality clusters and is independent of order of arrival of raw data.

All these details about diabetic patients will be useful features for the doctor to design apt treatment.

Table 3 Diabetic patients suffering from Parkinson

No. of clusters	Probability	Correlation
1	35	42
2	379	293
3	994	1758
4	712	27

CFBA [8, 9] is a statistical probability-based incremental clustering algorithm. CFBA proved best results in various domains of engineering, for incremental learning, knowledge augmentation and mapping. To enhance CFBA and to remove the probability approach, CBICA is developed and presented in this paper, as new correlation-based incremental clustering algorithm. Table 1 gives the details regarding opting for correlation as matrix for sampling population. Table 3 shows the exact results of Diabetic patients' datasets who are suffering from Parkinson. This data set is considered in this study to further add flavours in incremental learning and to understand the quality clusters details produced by CFBA and CBICA both. As shown in Table 3, the nature of cluster formation remains almost same in both approaches. But clusters formed by CBICA gives correlation (including positive, negative and zero) between data series, which is missing in probability-based approach. These correlation details of clusters will help support the 'mass customization' or personalization approach of diabetic treatment.

3.4 Cluster Analysis of CBICA for Multiple Iterations

- Figures 9 and 10 represent the average values of each attribute for each obtained cluster.
- These clusters are obtained from the multiple iterations of the CBICA, where the total instances are divided into two halves, that is, 7500 instances each, to obtain the multiple iterations.
- After analysing the figure, it is conclusive that TRIGLYCERIDES is the most impactful attribute and BLOOD GLUCOSE PP is the second most impactful attribute.
- Thus the impactful attributes are same in both single as well as multiple iterations, which shows that CBICA is consistent irrespective of the order of the influx of data.

Figure 11 shows Cluster average values versus Outlier average values of CBICA. As shown in Fig. 11 the overlapping attributes are the non-impactful attributes. Based on execution results and cluster formation/reformation (on arrival of new data), it was feasible to rank the clusters and redefine the concept called 'outlier clusters' as Maverick Cluster. Based on the obtained results of execution of diabetic patients

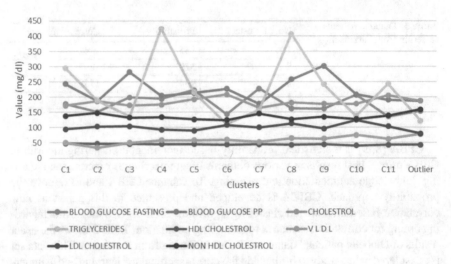

Fig. 9 Shows execution of CBICA using single iteration

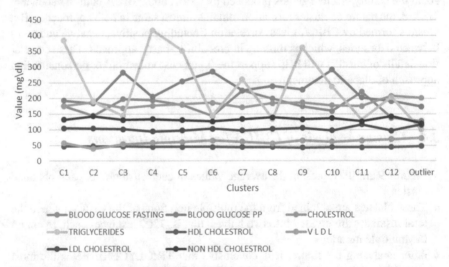

Fig. 10 Shows execution of CBICA using multiple iteration

data shown in Figs. 5, 6, 7 and Tables 2 and 4 here are the more details about cluster ranking and Maverick cluster.

- **Cluster Ranking**:

  – The most updated cluster has the highest cluster ranking.
  – Cluster C5 has the highest cluster ranking, as it is updated for the highest number of times in the multiple iterations.
  – Cluster C9 has the second highest cluster ranking.

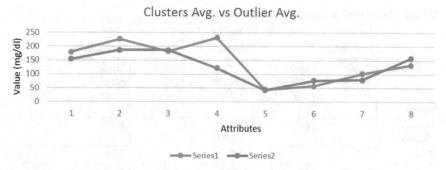

Fig. 11 Shows Cluster average values versus Outlier average values of CBICA

- Cluster C2 has the least cluster ranking, as it is updated for the least number of times in the multiple iterations.

- **Outlier Cluster: [Maverick Cluster]**

  - The cluster which does not get updated in the multiple iterations is termed as the Outlier Cluster or Maverick Cluster.
  - Where Maverick means an unorthodox or independent-minded or outsider.
  - Cluster number 13 was observed as the Outlier cluster, as it was not updated in the multiple iterations.

3.5 Validations of Results

In addition to testing the new algorithm CBICA for purity, order-independence, threshold matching, incremental clustering, etc. it was essential to apply intra and inter-cluster validations too. Following paragraph shows the details about further validations carried out related to CBICA:

- **Intra-Cluster Distance**:

  - The **Intra-Cluster distance** is the average value of the distances from the centroid of the respective cluster.
  - Lower the value, closer the cluster members, which indicates good compactness of the cluster.
  - Thus, from the results obtained, the intra-cluster distance value of the CBICA, for single iteration, is best with the least value of 0.077932.

- **Inter-Cluster Distance**:

  - The **Inter-Cluster distance** is the average value of the distances between the various cluster centroids.
  - Higher the value, the separation of clusters is good, which indicates that the objects that are not in the same cluster are far away from each other, which means the clusters are well separated.

Table 4 Comparison of incremental clustering algorithms

Variables	CFBA		CBICA		XMeans	
	Probability		Correlation		Using Weka	
	Single iteration	Multiple iteration	Single iteration	Multiple iteration	Single iteration	Multiple iteration
Total instances	15000	7500 + 7500	15000	7500 + 7500	15000	7500 + 7500
No. of clusters	6	8	12	13	5	5
Purity	81%		87%		80%	
Entropy	2.1732	2.5921	2.9073	3.0777	2.3214	
Intra-cluster distance	0.301145	3.466204	0.077932	0.098994	–	
Inter-cluster distance	21.24176	52.42456	0.318746	0.362081	–	
Impactful attributes	Triglyceride, Blood Glucose PP, Blood Glucose Fasting	Triglyceride, Blood Glucose PP, Blood Glucose Fasting	Triglyceride, Blood Glucose PP	Triglyceride, Blood Glucose PP	VLDL, Cholesterol, Blood Glucose PP	VLDL, Cholesterol, Blood Glucose PP

– Thus, from the results obtained, the inter-cluster distance value of CFBA, for multiple iterations, is best with the highest value of 52.42456.

Non-Probability Sampling concept:

- An Internet-based research study shows that there are 12 types of Non-Probability Sampling.
- Out of which, this research mapped Correlation-based incremental clustering algorithm concept to the Total Population Sampling and Purposive sampling.
- Surveying every single member of a qualifying subgroup is obtained using CBICA.
- As failure to survey even a few respondents can make it difficult to generalize.

There are several different types of blood glucose tests, as authored by David et al. (2016)

- The first pathological test done to check for prediabetes and glucose levels is called Fasting Blood Sugar (FBS). This test is conducted after 8 h of individual's have taken last meal.
- Another type of test is taken after exactly 2 h of start of eating meal, termed as 2-h postprandial blood sugar measure. This is not the test to diagnose diabetes. This test indicates whether a diabetic patient is consuming correct amount of insulin with meals or not.
- Regardless of when an individual last ate the meal, Random Blood Sugar (RBS) measures blood glucose, can be taken throughout the day. Random testing is useful

Patient's name	Age	RBS	Closeness value
MRS. SARITA	35	110	0.210676
MR. RAM PR.	35	113	0.63408
MR. NAVEEN	35	132	0.048659
MR. RAM PR.	35	142	0.170319
MRS.ARUNA	35	142	0.740385
MR. R.PAUL	35	144	0.10902
MR. VISHNU	35	144	0.843887
MRS. RAMKA	35	157	0.035724
MRS. ANIL	35	162	0.271227
MRS. SHANT	35	181	0.047505
MRS. JASBIR	35	185	0.436804
MR. RAVI	35	191	0.438745
MR. PRAMOI	35	196	0.970066
MRS. REENA	35	199	0.114984
MR. RAM PR.	35	200	0.173149
MRS.ARUNA	35	200	0.943063

Patient's name	Age	RBS	Closeness value	Patient's name	Age	RBS	Closeness value
MR. BAL	36	205	0.32446	MS. PRA	40	172	0.916098
MR. BAL	36	190	0.97919	MS. PRA	42	204	0.445935
MR. BAL	38	126	0.470924	MS. PRA	43	197	0.957006
MR. BAL	38	220	0.863185	MS. PRA	43	158	0.780926
MR. BAL	40	145	0.237814	MS. PRA	49	150	0.48729
MR. BAL	41	200	0.766264	MS. PRA	49	208	0.694189
MR. BAL	44	167	0.415736	MS. PRA	49	198	0.593046
MR. BAL	46	196	0.856649	MS. PRA	50	220	0.951725
MR. BAL	47	155	0.489755	MS. PRA	53	205	0.579105

Fig. 12 a Shows the computations performed on various patients based on their age, RBS and closeness values, **b** shows male individual details over the period of 36–47, **c** shows female individual details over the period of 40–53

because glucose levels in healthy people do not vary widely throughout the day. Blood glucose levels that vary widely indicate a problem.

- Oral Glucose Tolerance Test is conducted to diagnose diabetes that occurs during pregnancy (gestational diabetes). Women who had high blood sugar levels during pregnancy may have oral glucose tolerance tests after pregnancy. After pregnancy continuous glucose monitoring in mother and child may lead to a new research indicating mapping between gestational to juvenile to type 1/2 diabetes.
- To know how much sugar is stuck in red blood cells, Haemoglobin A1c is carried out. This test can be used to diagnose diabetes. It also shows how well your diabetes has been controlled in the past 2–3 months and whether your diabetes medicine needs to be changed. The result of your A1c test can be used to estimate your average blood sugar level. This is called your estimated average glucose, or eAG. Such tests data is very essentially used in modelling personalized systems [21, 22] (Fig. 12).

Three hundred and thirty two patients clinical data is studied over for this research, and each patients 10–20 repetitive exam reports were studied for analysing personalization details. These patients include male and female both, having varied ranges of blood glucose. The analysis was carried out using Cobweb clustering algorithm.

Table 1 shows the analysis of various patients based on RBS values and Cobweb clustering result for a specific age = 35. This table contains both male and female patients' details, with variation in closeness values, due to the range of RBS values.

Table 2 shows the analysis of a patient Mr. Bala, over the period of time from age = 36 to age = 47. Closeness values are inversely proportional to RBS values.

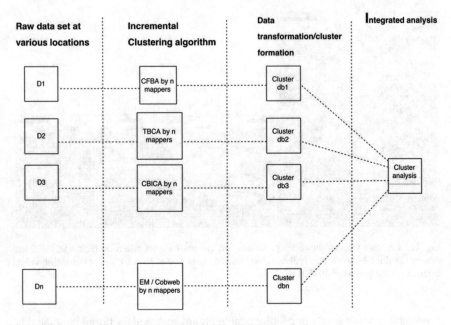

Fig. 13 Shows distributed proposed framework of CBICA

Table 3 is an extension of Table 2, wherein female patients details are visible from age = 40 to age = 53. It is visible that closeness values are differently proportional to RBS (inversely proportional but in different manner as against male candidates).

From all these tables and results it is visible that Personalization related to various blood glucose levels of individual patients, its empirical data analysis is a must for every prescription. Diabetologist need to consider the report generated by IT-based system related to individual patient's details (including diet, exercise regime, family-history, course of medicines, age, gender, etc.). The personalization of this kind will not only suggest apt treatment to patient and may maintain/reduce blood glucose levels but also reduce per-patient treatment time of diabetologist and provide outcomes useful for future researchers.

As shown in Fig. 13, CBICA is transforming into the complete distributed system of its own. CBICA is deployed on MapReduce platform using various mappers, reducers and combiners. In this research it is also planned that distributed CBICA can make automatic selection of number of mappers and reducers. This distributed system is under construction and will have two scenarios to implement. First scenario will have centralized server having implementation of CBICA. All connected nodes will send the diabetes datasets to server/CBICA to execute and send results/clusters back to nodes for further analysis. In second case, implementation of CBICA will be available with every node. Hence clustering can be achieved at every node, with further analysis as and when needed. Server will combine the results and analysis as per the need.

4 Conclusions, Summary and Future Directions

This paper helps to propose a system which overcomes the drawbacks of the existing incremental clustering algorithms, which uses the probability-based similarity measures. The proposed system uses non-probability-based similarity measure, i.e. Pearson's coefficient of correlation, and shows promising results over the probability-based incremental clustering algorithm. As the range of correlation is between -1 to $+1$, the outcomes suggest whether the two data series are negatively correlated or positively correlated. This is an added advantage over the probability-based incremental clustering algorithm. The proposed system calculates the threshold values by taking the average of all the closeness values, which are used to assign the input data to the desired clusters.

CBICA's detection capabilities of high impact attributes can prove to be the key for treatment and prevention. This paper shows how incremental clustering is used to model actual diagnosis of diabetes for local and systematic treatment, along with presenting related work in the field. Experimental results show the effectiveness of the CBICA-based model.

This paper collects and analyses medical records of an individual diabetic patient with knowledge discovery techniques to extract the information from patient volunteering for check-ups. The experiment has successfully performed with data mining algorithm CBICA. Personalization for analysis of the medical of reports of the person will definitely be beneficial for the patient as well as doctor, by saving their valuable time. Personalization will help to get the accurate history of the patient to get the effective treatment. The outcomes of this experiment suggest that the increase in the diabetes level depends upon the impactful attributes which were obtained from the results. The outcomes will help the doctors to specifically suggest the treatment and the preventive measures to be taken by the patients. This project can further be optimized and later its focus can shift on increasing the number of reports as well as the attributes, thus more detailed hidden relationships can be found between the various attributes.

The current system can be further enhanced and expanded by incorporating with other clustering algorithms and can be designed in such a way that the algorithm becomes more accurate. In future, information from different locales over the globe can be gathered and a more precise model could be developed for diabetes parameter trigger prediction.

Further study will mainly focus on using this system for the incremental clustering algorithm of the various available domains including juvenile diabetes, gestational diabetes and other chronic diseases, to get more accurate results.

It is also necessary to incorporate cluster ranking in unique way, to further extend this study. Clusters need to be ranked high, if they are highly updated with every iteration of influx of new data. The principal component in that cluster remains constant throughout all iterations, should not change the ranking of attributes with every iteration. The most valuable attribute need to be noted which is used for decision making or predictive analysis by authorities of an organization. 'Representative of the cluster' is the major source of incremental clustering as this member is compared with every iteration for appending cluster or forming new cluster. It is necessary to check what affects the changes in this centre of cluster, related to changes in threshold value, per iteration. Can we also introduce two representatives of clusters? Naming the clusters should happen automatically by looking at varying values of impactful and highly ranked attributed of every cluster. Such naming will be useful for authorities or researchers to utilize a specific cluster based on need.

References

1. Aghabozorgi, S., Wah, T.Y.: Effective clustering of time-series data using FCM. Int. J. Mach. Learn. Comput. **4**(2), 170 (2014)
2. Bagnall, A., Lines, J., Bostrom, A., Large, J., Keogh, E.: The great time series classification bake off: a review and experimental evaluation of recent algorithmic advances. Data Min. Knowl. Discov. 1–55 (2016)
3. Bertsimas, D., Kallus, N., Weinstein, A.M., Zhuo, Y.D.: Personalized diabetes management using electronic medical records. Diabetes Care **40**(2), 210–217. https://doi.org/10.2337/dc16-0826. http://care.diabetesjournals.org/content/40/2/210
4. Cheryl, A., Lidong, W.: Big data in healthcare: a new frontier in personalized medicine. Med-Crave Open Access J. Transl. Med. Res. http://medcraveonline.com/OAJTMR/OAJTMR-01-00005.pdf (2017)
5. Frank, M., Streich, A.P., Basin, D., Buhmann, J.M.: Multi-assignment clustering for boolean data. J. Mach. Learn. Res. **13**, 459–489 (2012)
6. Khaleghi, A., Ryabko, D., Mary, J., Preux, P.: Consistent algorithms for clustering time series. J. Mach. Learn. Res. **17**(3), 1–32 (2016)
7. Kulkarni, P.A., Mulay, P.: Evolve systems using incremental clustering approach. Evol. Syst. **4**(2), 71–85 (2013)
8. Mulay, P., Kulkarni, P.A.: Knowledge augmentation via incremental clustering: new technology for effective knowledge management. Int. J. Bus. Inf. Syst. **12**(1), 68–87 (2013)
9. Mulay, P.: Threshold computation to discover cluster structure: a new approach. Int. J. Electr. Comput. Eng. **6**(1), 275 (2016)
10. Nanda, S.J., Panda, G.: Design of computationally efficient density-based clustering algorithms. Data Knowl. Eng. **95**, 23–38 (2015)
11. Shinde, K., Mulay, P.: CBICA: correlation based incremental clustering algorithm. Int. J. Control Theory Appl. **10**(9), 575–582 (2017)
12. Sharon, T.: Self-tracking for health and the quantified self: re-articulating autonomy, solidarity, and authenticity in an age of personalized healthcare. Philos. Technol.
13. Pritchard, D.E., Moeckel, F., Villa, M.S., Housman, L.T., McCarty, C.A., McLeod, H.L.: Strategies for integrating personalized medicine into healthcare practice. Personal. Med. **14**(2), Research Articlefree. https://doi.org/10.2217/pme-2016-0064. https://www.futuremedicine.com/doi/full/10.2217/pme-2016-0064. Accessed 19 Jan 2017

Web References

14. ADA Dailyon: More personalized approach to type 2 diabetes care can help prevent overtreatment, June 10, 2017, http://www.adadaily.org/2017/06/10/more-personalized-approach-to-type-2-diabetes-care-can-help-prevent-overtreatment/

15. dia Diabetes Atlas. http://www.diabetesatlas.org/ (2015). Accessed 03 Apr 2017
16. New Research Further Validates Benefits of Personalized Diabetes Therapy Management Across the Continuum of Care. https://www.glytecsystems.com/News/new-research-further-validates-benefits-of-personalized-diabetes-therapy-management-across-the-continuum-of-care.html
17. Malay, G., Teresa, W.: The future of personalized healthcare: predictive analytics. Rock Health 2017. https://rockhealth.com/reports/predictive-analytics/
18. Phillips, K.A., Sakowski, J.A., Liang, S.Y., Ponce, N.A.: Economic perspectives on personalized health care and prevention. Forum Health Econ. Policy 16(2), 57–86, 01 Sept 2013. https://www.sutterhealth.org/research/publications/economics-personalized-health-care-prevention
19. Narain, S.: Why 'standardized' and 'personalized' care are two sides of the same coin. The Bridge, March 6, 2017. https://www.advisory.com/research/health-care-industry-committee/the-bridge/2017/03/standardization-and-personalization
20. Pane, J.F., Steiner, E.D., Baird, M.D., Hamilton, L.S., Pane, J.D.: Insights on personalized learning implementation and effects. Informing Progress, Rand Education, July 2017. https://www.rand.org/pubs/research_reports/RR2042.html
21. Derrer, D.T., MD: "Do Fasting Diets Work?", WebMD Medical Reference Reviewed by on March 19, 2016
22. Vinnie Ramesh, Chief Technology Officer, Co-founder of Wellframe: The Future of Personalized Healthcare: Predictive Analytics (2016)

Processing Using Spark—A Potent of BD Technology

M. Venkatesh Saravanakumar and Sabibullah Mohamed Hanifa

Abstract Processing, accessing, analyzing, securing, and stockpiling of big data are the most core modalities in big data technology, where **Spark**, is a core processing layer, an open-source cluster (in-memory) computing platform, unified data processing engine, faster and reliable in a cutting-edge analysis for all types of data. It has a potent to join different datasets across multiple disparate data sources. It supports in-memory computing and enables faster query access compared to disk-based engines like Hadoop. Query ID="Q1" Text="Please check and confirm if the author names and initials are correct." This chapter sustains the major potent of processing behind Spark connected contents like Resilient Distributed Datasets (RDDs), scalable Machine Learning libraries (MLlib), Spark incremental Streaming pipeline process, parallel graph computation interface through GraphX, SQL Data frames, SparkSQL (Data processing paradigm supports columnar storage), and Recommendation systems with MlLib. All libraries operate on RDDs as the data abstraction is very easy to compose with any applications. RDDs are a fault-tolerant computing engine (RDDs are the major abstraction and provide explicit support for data-sharing (user's computations), can capture a wide range of processing workloads and parallel manipulated can be done in the cluster as a fault-tolerant manner). These are exposed through functional programming APIs (or BD-supported languages) like Scala, Python. Chapter also throws the viewpoint on core scalability of Spark to build high-level data processing libraries for the future generation application is involved. To understand and simplify the entire BD tasks, focusing of processing hindsight, insights, foresights by using Spark's core engine, its members of ecosystem components are explained with a neat interpretable way, is mandatory for data science compilers at this moment. Big contents dive (current big data tools in Spark, cloud storage) of cognizance are explored in this initiative to replace the bottlenecks towards the development of an efficient and comprehend analytics applications.

M. Venkatesh Saravanakumar (✉) · S. M. Hanifa
Research Department of Computer Science, Sudharsan College of Arts & Science,
Pudukkottai 622104, Tamil Nadu, India
e-mail: venkatesh.srivi@gmail.com

S. M. Hanifa
e-mail: manavaisafi@yahoo.com

1 Escalation—Apache Spark

Spark, cluster computing architectural model to Big data solution, works both in-memory and on disk since it holds intermediate results in memory instead writing data on disk. (to work on a same processing dataset in a multiple times). Organizations that are new to Hadoop will typically begin their works with spark, is the most active open-source project in big data and has a caliber to replace Map Reduce (MR) (Need not write MR code). It is a flexible in-memory data processing for Hadoop in the line of easy development, flexible, extensible API and fast batch and stream processing. Performance is faster in in-memory data storage and near and real-time processing (like spark) compared with other big data technologies. Supporting of lazy evaluation of big data-queries, it helps in optimization of the data processing workflows. Also, provides a high-level APIs (Computing Interface) to improve productivity and development of architect model for big data solutions.

1.1 History of Apache Spark

Initially Spark was developed as research project [1]. In course of time researchers found that Map Reduce was not suitable for the iterative and interactive computing jobs. To overcome this, Spark was designed to work fully in-memory, so as to face interactive queries and iterative algorithms. Later periods, the additional and higher level components like Shark, Spark streaming was developed. Initially Spark was developed and later transferred to the Apache Software Foundation. Spark has become one of the largest open-source communities in Big Data. Now, more than 250 contributors in over 50 organizations are contributing to Spark development. The user base has augmented tremendously from small companies to Fortune 500 companies.

1.2 Spark Architecture—Core Components

A core component of Spark extends task dispatching, scheduling and Input and Output distribution functionalities. It has the architecture of

- Data Storage (HDFS, Hbase, Cassandra etc.,)
- API (Computing Interface)

  - Structured API (Data Sets, Data Frames, SQL)
  - Low Level APIs (Distributed Variables, RDDs)

- Management Framework-Resource Management.

Table 1 Big data supported API languages

Language APIs	Spark Stream	Storm	Triden	Samza	Apache Flink
	Java, Scala, Python	Java, Scala, Python, R, Clojure	Java, Scala, Python	Java, Scala	Java, Scala

1.3 Creating Spark Applications—BD Languages (APIs)

Spark runs on Java Virtual Machine (JVM) environment and developed in Scala programming language. Big data supported languages like

- Scala
- Java
- Python
- Ruby
- Clojure and
- R

These (Listed in Table 1) can be used to develop Big Data dependent applications.

1.4 Spark Key-in Features Are

- Supports MR functions.
- Workflows in Directed Acyclic Graph (DAG).
- Improved and optimized workflow.
- Supports Computing interfaces in Scala, Java and Python.
- Provides communicative shell.

1.5 File System Supported by Spark

The three file systems are being supported by spark [2] are named as

- HDFS (Hadoop Distributed File System)
- Local File System (LFS)
- S3 (Amazon Web Server (AWS) Cloud) and S4.

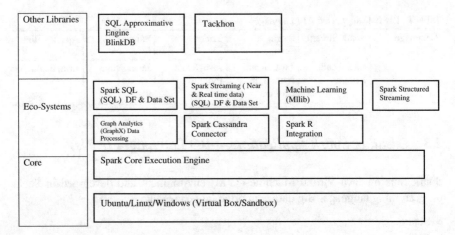

Fig. 1 Apache spark ecosystem

2 Spark Ecosystem

Spark Streaming is an extension of core Spark API. It makes it simple to build fault-tolerant processing of real-time data streams. Spark supports Big Data analytics and Machine Learning areas only because of its strong Spark Core API, additional libraries. All are (Spark SQL—[Data Frames and Data Sets], Spark Streaming [Near and Real-Time data processing], ML, Graph Analytics [GraphX] data processing, Spark Cassandra Connector, SparkR Integration) the part of the Spark ecosystem (Illustrated in Fig. 1) along with Spark frameworks core engine.

2.1 Spark Streaming, Spark SQL, Spark MLlib, SparkGraphX, and SparkR

2.1.1 Spark Streaming

To process real-time streaming data, data fed through streaming input sources like Kafka, Flume, Twitter, ZeroMQ, Kinesis, or TCP sockets. DStream, is used for real-time processing and supported frameworks (explained in Fig. 2) are

- Apache Samza
- Storm and
- Spark Streaming.

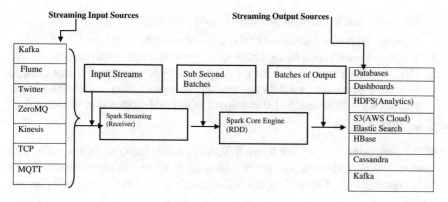

Fig. 2 Spark streaming

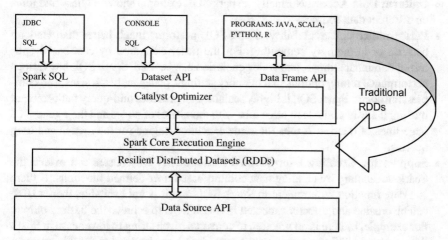

Fig. 3 Spark SQL

2.1.2 Spark SQL—Relational BD Processing

SparkSQL (see Fig. 3) [3] is defined as a query structured data that fits with a distributed dataset like schemaRDD/dataframe in Spark. It runs HiveQL queries, SQL. Some of the features are integrated (RDD with Spark SQL queries), unified data access (query sources—Hive tables, JSON), Hive compatibility and database connectivity (ODBC/JDBC).

Spark SQL History a new project called Shark was introduced into the Spark ecosystem in early versions of Spark. Shark used Spark as an execution engine instead of the Map Reduce engine for executing hive queries and generates an abstract syntax tree, converted to a logical plan with some basic optimizations. It adopts optimization techniques and created a physical plan of RDD operations, then applies on Spark. Since it improve in-memory performance to Hive queries.

But, Shark had three major problems to deal with

1. Running relational queries on RDDs was not possible
2. Running Hive QL as a string within spark programs was error-prone
3. Hive optimizer was created for the Map Reduce paradigm and it was difficult to extend Spark for new data sources and new processing models. Shark was discontinued from version 1.0 and Spark SQL was introduced in version 1.0.

Spark SQL features

SQL Integration with Spark Spark SQL helps easily combine SQL queries with spark programs. Users can post query within the Spark environment via data frame API or through SQL. Below is a spark SQL example on how one can apply for getting results.

- **Uniform Data Access**: Normally, a spark SQL example shows a query and joins on different data sources.
- **Performance and Scalability**: Spark SQL performs much better than Hadoop because of in-memory computing. For the faster execution by cost based optimizer, columnar storage, and code generation can be used. Spark SQL has various performance tuning options like memory settings, codegen, batch sizes, and compression codes. Spark SQL is highly scalable and provides mid-query fault-tolerant making it easily scalable to large jobs with thousands of nodes and their generated huge lines of queries. It uses the same execution engine for interactive and long queries.
- **Support for Creating User-Defined Functions (UDF)**: Users can extend the Spark vocabulary by creating new column based user-defined functions. If there is a new function requirement in Spark SQL and it is not available, then a UDF can be created and directly executed in Spark SQL to achieve the desired output. For example, if there is no function to convert a given string to lower case in Spark SQL then an UDF (for example to Lower Case) can be created for this purpose. Every new UDF created must be registered with the list of functions.
- **Compatible with Hive**: Spark SQL reuses Hive meta store and front-end providing total compatibility with the execution of Hive queries and UDF's.
- **Provides Standard Connectivity through JDBC or ODBC driver**: The server mode in Spark SQL provides standard JDBC or ODBC connectivity for BI tools like Tableau.

Spark SQL Use cases it finds great application in Twitter Sentiment Analysis where organizations can identify trending topics to create promotional campaigns and attract larger audience. It helps in organizations to manage crisis, adjust their services and target marketing. Also it is used for detecting credit card frauds and any other fraudulent banking transaction. For instance, a credit card is being currently swiped in California and after 15 min the same credit card is being swiped in India, then there is a possibility of fraud which can be identified in real time using Spark SQL. Spark SQL also finds great application in the healthcare industry. One common use case of Spark SQL in the healthcare domain is for identifying genomic sequences. It is used to identify people who are predisposed to most of the common.

2.1.3 Spark MLlib—Library of Machine Learning Algorithms

Both common learning algorithms and utilities renders a ML Library called, MLlib. It includes the different sets of techniques as well as related algorithms.

2.1.4 Spark GraphX (Graph Analytics/Visual Analytics)

Graph Processing—Introduction

Definition—Graph is a collection of vertices connected to each other using edges. Vertex is a synonym for node, which can be a place or person with associated relationships expressed using edges. Just imagine expressing the whole of Facebook as a social graph, where finding relationships between them can be a very complex task.

In Apache Spark, creation of Directed Acyclic Graph (DAG) for Spark jobs that consist of Resilient Distributed Datasets *(RDD)* which are nothing but *vertices*, and *transformations are* nothing but *edges*. There are multiple ways to go from one place to another place. If we create places as vertices and roads as edges, graph processing can provide an optimized shortest path between these two places.

Two Technologies of Graph Processing

1. Graph Databases
2. Graph Processing systems

Graph Databases versus Graph Processing Systems there are two types of technologies in graph processing; graph databases and graph processing systems. Graph databases can be seen as OLTP databases that provide transactions, updates, deletes, and query language.

Graph processing systems can be seen as OLAP systems that provide offline analytic capabilities. The following table shows the important graph databases and graph processing systems:

Graph database systems are Neo4 J, Titan, OrientDB, AllegroGraph, GraphBase, Oracle Spatial, and Graph, NoSQL databases such as HBase and Cassandra are the examples and can also be used for storing graphs.

Graph processing systems are GraphX, Apache Giraph, GraphLab, Apache Hama, and Microsoft's Graph Engine. The GraphLab framework implements the Message Passing Interface (MPI) model to run complex graph algorithms using data in HDFS.

Google's Pregel project, a model which works on parallel processing, synchronizing, laid a foundation for Apache Giraph and Apache Hama. GraphX is built on top of Spark and supports a variant of the Pregel API as well.

Apache Giraph, born at LinkedIn, is a stable system that can process a trillion edges on top of Hadoop, but Apache Giraph is not supported by all major Hadoop vendors. If your use case is a pure graph-related problem, you can use Apache Giraph as a robust and stable solution, but if graph processing is just a part of the solution,

GraphX on top of Spark provides a great unified solution with the power of Spark core capabilities.

GraphX—Introduction—Spark's original Graph Analysis package is a graph processing system built on top of Apache Spark, which can be easily integrated with other Spark modules to unify ETL, exploratory analytics, and graph analytics. GraphX creates graphs as another type of RDD such as *Vertex RDD* and *Edge RDD*. Spark's GraphX features such as speed of iterative processing and *in-memory capabilities removed the issues of Map Reduce that Apache Giraph and Pregel were designed to address*. GraphX stores a graph's edges in one file and vertices in another file. This allows graph algorithms implemented in GraphX to view graphs either as graphs or as collections of edges or vertices to efficiently combine multiple processing types within the same program [3].

Graph Analytics

Operations of GraphX Spark-shell interactively work with GraphX. Vertex and edge files are called input data for GraphX operations and then store it on HDFS.

Creating a graph

1. Create a vertex file with vertex ID, name, and age.
2. Create an edge file with vertex ID, destination vertex ID, and relationship
3. Copy both files to HDFS:
4. Start a Scala shell with the master as YARN-client and then import GraphX and RDD dependencies. Note that GraphX is not supported in the Python language.
5. Create an RDD for both vertex and edge files:
6. Create the *VertexRDD* with *VertexId* and strings to represent the person's name and age:
7. Create the *EdgeRDD* with source and destination vertex IDs converted to Long values and the relationship as String. Each record in this RDD is now an Edge record.

It should be defined as a default value in case a connection or a vertex is missing. The graph is then constructed from these RDDs—vertices, edges, and the default record:

Counting

It is easy to count number of vertices and edges in a graph with a function

Filtering

Graphs can be filtered to create sub-graphs.

inDegrees, outDegrees, and degrees

GraphX also defines a special data structure for node degree as inDegrees, outDegrees, and degrees. The number of links it has with other nodes are denoted by degree of a node. The number of incoming links to a particular node is called incoming degree or an in-degree. Also, it is possible to find the outgoing degree, or out-degree, which is the number of nodes that it points to.

Transforming Graphs

Graph operators can be used to either change the properties of graph elements or modify the structure of graphs. To examine join methods, combine graph data with other datasets and perform data operations on *VertexRDD* and *EdgeRDD*.

Transforming attributes

The map is one of the main transformation functions for transforming RDDs in Spark. Similarly, graphs also have three map operators namely *mapVertices, mapEdges, mapTriplets*.

- For *mapVertices*, myMap takes a pair of (VertexId, VD) as input and returns a transformed vertex attribute of type VD2.
- For *mapEdges*, myMap takes an Edge object as input and returns a transformed edge attribute of type ED2.
- For *mapTriplets*, myMap takes an EdgeTriplet object as input and returns a transformed edge attribute of type ED2.

Modifying graphs

The GraphX library also provides four useful methods for changing the structure of graphs.

The operators used in the code are explained as follows:

1. **reverse**: If we want to reverse the edge directions, the reverse operator can be used. This does not cause any data movement and also does not change vertex or edge properties. This is useful for inverse PageRank as an example.
2. **subgraph**: The subgraph operator is useful for filtering graphs that take two predicate functions as arguments, which return Boolean values.
3. **mask**: The mask operator also filters a graph as in subgraph. But, mask does not take predicate functions as arguments.
4. **groupEdges**: It reduces the size of graph in many applications as group Edges operator which merges duplicate edges between each pair of nodes. It reduces the range of graph in many applications.

Joining graphs

GraphX also provides APIs for joining RDD datasets with graphs as well. This is really useful when it is needed to add extra information or merge vertex attributes to any vertex attributes.

VertexRDD and EdgeRDD operations

All of the graph operations are invoked on a graph and return a new graph object. Operations that transform *VertexRDD* and *EdgeRDD* are called collections of RDD subtypes.

Graph Algorithms here GraphX provides the following algorithms out-of-the-box in version 2.0. Some of the algorithms are:

- Connected components
- Label propagation
- PageRank
- SVD++
- Shortest Path
- Strongly connected components
  - Triangle count.

Triangle counting

To find the route, Triangle count algorithm can be used which works on the basis of counting vertex based Triangles.

Connected components

Unconnected graphs can be isolated from one another without affecting the others. When a graph is created. When a large graph is created from the data, it might contain unconnected sub-graphs that are isolated from each other and contain no connecting edges between them. This algorithm provides a measure of this connectivity. Depending upon our processing, it is important to know that all the vertices are connected. In two graph methods, connected Components and strongly Connected Components, can be used. Maximum iteration count is set to 1000, so as to call a strong method as these counts works on the graph vertices. Then the vertex counts are grouped with original vertex records, to achieve connection count, and this can be associated with vertex information, such as the person's name:

Case Study—Flight data using GraphX using Pregel API

With Pregel API, to analyze flight data airports are represented as vertices and routes as edges through basic graph analytics to find out departures and arrivals can be found and also analyze the data to find out the cheapest fares.

Many core graph algorithms are iterative algorithms, Pregel is an iterative graph processing model, developed at Google, which uses a sequence of iterations of message-passing between vertices in a graph. GraphX implements a Pregel for message-passing. In Pregel implementation, the vertices can only send messages to neighboring vertices.

Execution of Pregel operator

- Vertices receive inbound messages from the predecessor process
- Computation of new value for the vertex property
- Passing messages to the neighboring vertices

When there are no more messages remaining, the Pregel operator will end the iteration and the final graph is returned.

Spark Memory usage uses the following processes:

- RDD Storage
- Shuffle and Aggregation Purpose and
- User Code

RDD Storage Spark's storage levels provide different trade-offs between memory usage and CPU efficiency. Do not replicate the RDD storage until fast fault recovery is needed. If entire RDD fits in memory, choose MEMORY_ONLY as a process.

Shuffle and aggregation buffers Spark can create intermediate buffers while performing shuffle operations, later results may be aggregated.

User code arbitrary user code is executed by spark when user functions require substantial memory. Spark will comfortable for overall memory usage, when user application allocates a big arrays for processing. Only after allocating the space for JVM and RDD, space for the Used Code will be allocated.

2.1.5 SparkR

It invokes Spark into R, because it is single task process and lack of memory with Spark's distributed engine. It provides Data Frame in a distributed manner via MLlib to provide higher scalability. It invoked from shells, scripts, RStudio, as well as Zeppelin notebooks which also suitable for. It can be used with any platform and also can access YARN resource managers. By having additional API, data can be imported to SparkR.

Spark—Use cases
Apache Spark (Databricks) lists the following use cases:

- Data integration and ETL on MR
- Interactive analytics or BI
- Batch computations with high performance
- ML and advanced analytics
- Near and Real-time stream processing

Motivation behind Spark choice

Though Hadoop is in existence, more usage of Spark is evolving, thus provides more reasons like:

- Iterative Algorithm
- In-Memory Processing
- Near and real-time data processing and
- YARN (Yet Another Resource Allocator)

Iterative Algorithm: Map Reduce is not compatible with processing iterative algorithms.
In-Memory Processing: It stores and reads the data in nature.

Table 2 RDD Operations—*Transformations*

Manipulations (on single RDD)	Across RDDs on multiple RDDs)	Reorganization (on single RDD) (key-value)	Tuning (on single RDD)
Map	Union	groupByKey	Repartition
flatMap	Subtract	reduceByKey	Coalesce
Filter	Intersection	sortByKey	–
Distinct	Cartesian	–	–
–	Join	–	–

Near and real-time data processing: To handle streaming near and real -time data
YARN: It is a framework to deal with Cluster via pluggable scheduler.
MR Running: A higher level of operators and ML algorithms can be executed. Some
of the example of operators are; map(), filter(), reduceByKey(), groupByKey() etc.,

Creation of Programming Model

Resilient Distributed Dataset (RDD)—[4] It is a *Key programming abstractions*
for parallel programming. Programming model consists the following:

- RDD
- Parallel Operations and
- Shared Variables (Broadcast, Accumulators)

RDD [5] is a immutable, basic and key programming abstraction in Spark, illustrates elements partition and its parallel operation. RDDs are abstraction of memory
which is distributed in nature, works with the distributed manner by moving from
clusters to in-memory without making any errors.

RDD is a

- **Resilient** able to withstand and quickly recover from difficult conditions.
- **Distributed**: done across different node/clusters.
- **Dataset**: is a collection of divided/classified data.

RDDs parallel job processing it works in a parallel manner. RDDs are divided
naturally into the partitions and written in parallel, the data can be quickly accessed
in a sequential manner, as the case in Hadoop MapReduce framework.

Parallel Operations on RDD RDDs support two kinds of operations (See Tables 2
and 3) [6];

- **Transformations**: idle operations that revisit a different RDD.

**There are two transformation such as Wide transformation [data movement
across partitions]**—groupBykey(), reduceBykey() and Narrow Transformations
[No data movement]- filter(), map(), flatMap(), sample(), union().

- **Actions**: operations that activate computation and come with few values.

Table 3 RDD operations—*Actions*

Data Fetching	Aggregation	Output
Collect	Reduce	Foreach
Take(n)	Count	Foreachpartition
First	countByKey	saveAsTextFile
takeSample	–	saveAsSequenceFile
–	–	–

Example: Reduce(func), Collect(), Count(), First(), Taken(), countBykey(), Foreach(funk).

Pair RDDs in RDD (Key-Value Pairs), Spark extends a special transformations and actions, called Pair RDDs, used for processing and performing in the network. Pair RDDs have special transformation called *reduceByKey*, and a *join* transformation that can merge two RDDs together by grouping the elements in the same key.

It has

- Broadcast variables
- Accumulators

    - **Broadcast variables**: Programmer can create a "broadcast variable" which handles the object that *wraps the value*, is copied only once in each worker.
    - **Accumulators**: These are implemented using a different "serialization trick".

RDD New features—New Avatar in Spark these are to deliver the right applications.

- Immutable
- Partitioned
- Lazy evaluated
- Persistence an
- Fault Tolerance

Cluster/Resource Managers in Spark in Resource Management, the following are act as cluster managers in spark;

- Standalone Cluster Manager—Static partitioning
- Local—Static partitioning
- YARN—Dynamic Partitioning
- MESOS-Apache—Dynamic partitioning

YARN is the resource management framework for processing data in parallel manner for batch processing the shared data for the analytics. While HDFS provides scalable, fault-tolerant, and cost-efficient storage for big data, YARN provides resource management to clusters. Multiple frameworks are run on top of HDFS and YARN are run on Hadoop 2.0. YARN is like an operating system for Hadoop, which manages the cluster resources (CPU and Memory) efficiently. Applications such as MapReduce, Spark, and others request, YARN to allocate resources for their tasks. YARN allocates containers on nodes with the requested amount of RAM and virtual CPU from the total available on that node.

MESOS [7] is a cluster manager, can run analytic workloads and long-running services as well. (for example, Web applications or key-value stores) on a cluster.

Core technologies of ML on Spark is run on spark as detailed below

- Data ingestion, cleansing, and transformation [Wide and Narrow transformations are the two variants]
- Performance evaluation of regression models
- Parameters evaluation of both training and testing sets and
- Exploring the Model by using testing data set, is called Model training.

Ingestion of Data, Transformation and cleaning the data in ML Pipeline, to take the data for training the models, data can be ingested [8] in different ways like user activity data from surfer, accessing web APIs etc., To perform preprocessing of data (missing, incomplete),finding potential errors, combining different data sources and data aggregation. Once preprocessing is over, it is necessary to transform data into vector format, so as to fit with ML models. Spark's core API/SparkSQL engine performs cleansing, exploration, aggregation, and transformation by using Scala, Java, or Python libraries. Spark's library of MLlib offers models like linear models and DT regression models.

Performance Evaluation of Regression Models designed models can able to predict the target variables which are compared with true values so as to find metric levels. There are some familiar performance evaluation metrics like Mean Squared Error (MSE), Root Mean Squared Error (RMSE), Mean Absolute Error (MAE), R-squared coefficient to evaluate the models..

Evaluation of training and testing sets parameters creation of training and testing datasets must fulfill the norms of 80:20% from the original dataset respectively. N-fold validation formula adopted when we have the shortfall in sample collection in a particular dataset.

Model Development through Proper Training (To test speed of ML) create a model, a Stochastic Gradient Descent (SGD) is used to train the model, and can be imported from the MLlib library.

ML Algorithms ML is a field of computer science, studies the computational methods that learn from data source. Some frequently used and efficient ML Techniques (methods) where supported algorithms (See Table 4) (essential parts of BDA) can be found under the categories of Classification, Regression, Clustering, Association (Recommendations), Dimensionality Reduction, Bayesian methods, Methods on Collaborative filtering. MLlib consists of supported algorithms for faster execution of programs. These can be categorized as

- Predictive (Supervised) Learning, i.e., Learning with labels

Table 4 ML categories, tasks and supported algorithms

Types of category (in ML)	Method (s)/task (s)	Supported algorithm (s)
Supervised [making predictions]	Classification (categorical–qualitative) regression (continuous-quantitative)	Linear SVM, NN, Naïve Bayes (NB), Ensembles of Trees (Random Forest (RF)), Gradient-Boosted Trees, DT. Ridge, Logistic, Linear (Regression), Survival Analysis, SVM
Unsupervised [extracting structure]	Segmentation/clustering (categorical–qualitative)	K-Means, Spectral Clustering, Gaussian Mixture, LDA
	Dimensionality reduction/decomposition (continuous-quantitative)	NMF, ICA, Elastic Net, Singular Value Decomposition (SVD), Principal Component Analysis (PCA)
Recommender/ recoommendations [associating user item]	Collaborative filtering (CF) (categorical–qualitative)	ARM, FP Growth, Item-Based CF, Alternative Least Squares (ALS) algorithm
Feature extraction (processing text—text mining)	Feature extraction/transformation (categorical–qualitative)	LDA, Tokenization
Optimization (finding minima)	Finding minima (continuous-quantitative)	Stochastic Gradient Descent,Limited-Memory (L-BFGS)
Input/output formats	Supports I/O formats	Support for LIBSVM, Data Integration via SQL and PMML, MLlib's internal format for model export
Primitives and utilities	Low level primitives and basic utilities	Convex Optimization, Distributed Linear algebra, Statistical analysis
Documentation	MLlib user guide	Provide Extensive Documentation, Describes all methods, utilities and API codes
Code dependencies	Open source libraries	Breeze, Netlib-java, NumPy

- Descriptive (Unsupervised) Learning, i.e., Learning without labels
- Semi-Supervised
- Hybrid or Deep Learning
- Counsel System
- Extraction of Features (Feature selection) and Transformation
- Dimensionality Reduction (unsupervised learning)
- Association and Regression methods and
- Methods on Collaborative filtering

Supervised learning deals with labeled training data. For example, historical e-mail training data will have e-mails marked as ham or spam. This data is used to train a model that can predict and classify future e-mails as ham or spam. Supervised learning problems can be broadly categorized into two major types—classification and regression

1. **Classification** predicts categorical variables or classes. A couple of examples are spam detection and predicting customer churn. This target variable is discrete and has a predefined set of values. The classification algorithms are as follows:

   - **Naive Bayes**: This algorithm makes predictions based on the conditional probability distribution of a label given an observation. This assumes that features are mutually independent of each other.
   - **Decision Trees**: This algorithm uses a decision tree as a predictive model, which maps observations about an item.
   - **Ensembles of trees** (**Random Forests** and **Gradient-Boosted Trees**):
   - Ensemble algorithms combine base decision tree models in order to build a robust model. They are intuitive and very successful for classification and regression tasks.

Regression: It deals with a target variable and is continuous. For example, to predict house prices, the target variable price is continuous and doesn't have a predefined set of values. The regression algorithms are as follows;

Regression Models (Linear Regression, **Logistic Regression**, and **Support Vector Machines**): Regression algorithms are expressed as convex optimization problems aiming to minimize an objective function based on a vector of weight variables. An objective function controls the complexity of the model through the regularized part of the function, and the error of the model through the loss part of the function.

Unsupervised learning (Clustering) dealt with unlabeled data. It aims to observe structure the data and discover patterns. Cluster analysis, association rule mining, outlier detection, dimensionality reduction are the tasks behind it for modeling.

Clustering Algorithms

- **K-Means**: This is the task of grouping similar objects (called a cluster) together to partition n observations into k clusters, for example, grouping similar customers together to target them separately, detecting abnormal
- data, and clustering of text documents.
- **Gaussian Mixture**: This is a probabilistic model that is also used for data clustering such as k-means.
- **Power Iteration Clustering** (**PIC**): This algorithm groups vertices of a graph based on pair wise edge similarities.
- **Latent Dirichlet Allocation** (**LDA**): This algorithm is used to group collections of text documents into topics.
- **Streaming K-Means**: This algorithm clusters streaming data dynamically using a windowing function on the incoming data. This is a really useful algorithm in Spark Streaming applications.

- **Dimensionality Reduction**: These algorithms aim to reduce the number of features under consideration. This reduces noise in the data and focuses on key features. This type of algorithms include the following:
- **Singular Value Decomposition** (**SVD**): This algorithm breaks the matrix that contains the data into simpler meaningful pieces. It factorizes the initial matrix into three matrices.
- **Principal Component Analysis** (**PCA**): This algorithm approximates a high dimensional dataset with a low-dimensional subspace.

Recommender systems (Counsel Systems) are used to recommend products or information to users. Examples are video recommendations on YouTube or Netflix.

Collaborative filtering forms the basis for recommender systems. It creates a user-item association matrix and aims to fill the gaps. Based on other users and items, along with their ratings, it recommends an item that the target user has no ratings for. Algorithm implemented in collaborative filtering are classified into three categories namely,

- User-based collaborative filtering
- Item-based collaborative filtering
- Matrix factorization with alternating least squares, SVD++, Parallel SGD

In Spark, one of the most useful algorithms is Alternating Least Squares, which is described as follows:

- **Alternating Least Squares** (**ALS**): ALS models is for minimizing the errors in the observation. Input data can be of two types—explicit feedback or implicit feedback from users. In explicit feedback, the relationship between a set of user-item pairs is directly known, for example, the presence of movie ratings from different users on movies. In implicit feedback, the relationship does not exist directly.

A recommender system has to infer user preferences from the presence or absence of movies watched or not, purchases, and clicks or search events. An implicit feedback problem is much harder than an explicit feedback problem.

Uses of Machine Learning is listed below

- Spam detection
- Voice recognition
- Medicine and Healthcare
- Advertising
- IoT (Internet of Things)
- Robotics
- Gaming
- Stock Trading and
- Retail and e-commerce.

Building ML Pipeline makes easy to compose a compound algorithms into a single pipeline. ML pipelines are constructed from the Data frames-API, which are built on top of Spark SQL. Concepts used in ML pipelines are

- **Data Frame**: It is used to create rows and columns of data like in RDBMS table. Data Frame contains text, attribute vectors, exact labels, and forecasting..
- **Transformer**: It is an algorithm to transform a Data Frame into another Data Frame with predictions. The ML model is the best example of a Transformer.
- **Estimator**: It is an algorithm to generate a Transformer by fitting on a Data Frame. Generating a model is an example of concept of Estimator.
- **Pipeline**: As the name indicates, a pipeline creates a workflow by chaining multiple Transformers and Estimators together.
- **Parameter**: It is an API to specify parameters.

Building Recommender/Counsel Systems/Engine is easy to build any recommender/counsel systems or engines through Spark and some of well-known examples of recommender systems are named here

- NetFlix/YouTube—Movie/Video Recommendations
- Amazon.com—Bought Section
- Sportify—Music Recommendations
- Google—News Recommendations

Content—Based Filtering systems build recommenders based on item attributes. Examples of item attributes in movies are the generic, actor, director, producer, and hero. A user's taste identifies the values and weights for an attribute, which are provided as an input to the recommender system. This technique is purely domain-specific and the same algorithm cannot be used to recommend other types of products. **Collaborative Filtering** systems recommend items based on similarity measures between items or users. Items preferred by similar users will be recommended to another user. Collaborative filtering is similar to how you get recommendations from your friends. This technique is not domain-specific and does not know anything about the items, and the same algorithm can be used for any type of product such as movies, music, books, and so on. User-based and item-based filtering are the two kinds of collaborative filtering.

- **User-based collaborative filtering**

User-based recommender system is based on the similarities between
Users. The idea behind this algorithm is that similar users share similar
preferences.

- **Item-based collaborative filtering**

Item-based recommendation is based on similarities between items. The idea behind this algorithm is that a user will have a similar preference for similar items. It is possible to combine content-based filtering and collaborative filtering to achieve optimized results.

Limitations of Recommendations System are as below

- The Cold-Start Problem
- Scalability

- The Lack of Right Data

Shortage behind a precise data in some situations, source of Input data not be accurate always, because human beings are not perfect at providing ratings. User behavior is more important than ratings. Item-based recommendations provide a better answer in this case.

Recommendation/Counsel System with MLlib while building a counsel system, it must facilitate the following stages to reach out the goal of BD architect model development with the help of MLlib are detailed below

- Environment Preparation (Platforms/tools/methods)
- RDD creation
- Presenting data with Data Frames
- Partitioning training and testing data from original datasets (80:20 ratio)
- Exploring a Model
- Visualize Predictions
- Perform evaluation model through testing data and
- Validating the results accuracy to derive a build a better model

Spark Applications are used for creating and scheduling large-scale data processing applications in production, while the spark-shell is used for development and testing. Applications can be created in natively supported languages such as Python, Scala, Java, Ruby, SQL, R, or external programs using the pipe method. *Spark-submit* are used for submitting a spark application.

2.2 Cloud Computing—Promising Technology in Storage

With emergence and huge transformational potential capability of emerging big data storage architectures—paradigm revealed a study by [9]. The influence and utilization on big data escalates rapidly, as cloud computing storage is also concurrently augmented with a high impact to build efficient cloud systems. It is a rapid and massive-scale and intricate computing which eradicates costly computing hardware, volume of memory space, and software. It has *front-end* (which includes computers and software's to access the cloud network) and *back-end* (*different*, servers, and databases) that makes the cloud. SaaS, PaaS, IaaS, and DaaS are the dominant cloud delivery models. In particular, cloud storage can be accessed everywhere from every device and is used by enterprises and end-users as well.

Role of Cloud Computing for Big Data (BD) abut of Cloud Computing and Big Data prototype is evolved to tackle the data dependent privacy and different levels of security challenges through BD driven cloud security [10]. It provides a skill of service to compute and manipulate the data sets via distributed queries. Cloud computing extends the core engine by employing the utilization of Spark core engine and ecosystems.

Cloud Deployment model consists four types are listed as below

1. Public cloud (off-site network)
2. Private cloud (On-site network)
3. Hybrid cloud and
4. Community cloud.

2.2.1 Apache Sentry

Though it is in incubation level, it strongly provides security and authorization provisions behind big data applications to meet the challenges. It is designed to be a pluggable authorization engine for Hadoop components. It allows authorization rules to validate a user or application's access requests for Hadoop resources. Secured big data through Apache Sentry can generate useful predictions in the domain of Smart health—potential applications [11] to preserve in the long term storage process (Maintaining patients' population and hospital data). Spark performs data analysis very rapidly even though data resides in multiple cores [12].

2.3 Summary

Big Data (BD), an emerging technology provides innovative information, methodology in all aspects of Data Science. It shows demanding and forceful task due to the increased demand in exploring Big Data, called Big Data Analytics (BDA). Machine Learning (ML), one of core component of data analytics explores the data-intensiveness and provides various interpretations, by learning from data. A new horizon in BD is developed, called *Spark*, a Core and ecosystem architecture, can effectively handle and explore the rapid evolution of BD, and generates useful predictions in the domain of Smart health. Moreover recommendation systems using MLlib, promising technology of storage like cloud computing and cloud security via Apache Sentry (supports security and authorization provisioning) exploration are also detailed in a grasping level. Scalability and security in large-real-time-streaming job approaches is nowadays essential in all fields, especially developing analytic models in health applications by using open-sourced cluster computing processing engine, called *Spark*. In a nutshell, over all coverage on spark in & outs and growing components are visualized.

2.4 Conclusion

Spark, largest open-source communities in big data, its architectural model improves productivity and application development to big data solution works both in in-memory and on disk. Big Data processing Technologies/Tools utilizes various applications domains through BD platforms like Spark are very much impacted by enterprise vendors and end-users in the *social science frontier revolution*. Hence, Spark usage is inevitable to extract accurate values from big data solution in the years to come.

References

1. Ankam, V.: Big Data Analytics. Packt Publishing (2016). ISBN 978-1-78588-469-6
2. Apache Spark Research. https://spark.apache.org/research.html (2014)
3. Ambrusty, M., et al.: Spark SQL: Relational Data Processing in Spark. AMP, UC Berkrley, (2015)
4. Lu, X., et al.: Accelerating spark with RDMA for big data processing: early experiences. In: 22nd annual Symbosium on High-Performance Interconnects. IEEE (2014)
5. Zaharia, M., et al.: Spark: Cluster Computing with Working Sets, Hot Cloud (2010)
6. Zaharia, M., et al.: Resilient distributed datasets: a fault—tolerant abstraction for in-memory cluster computing. In: NSDI' 12 USENIX Symposium on networked design and implementation with ACM SIGOCOMM and ACM SIGOPS, SAN-JOSE,CA (2012)
7. Hindman, B., et al.: Mesos: A Platform for fine-grained resource sharing in the data center, Technical report UCB/EECS-2010-87, EECS Department, University of California, Berkely, May 2010
8. Fu, J., et al.: SPARK—a big data processing platform for machine learning. In: International Conference on Industrial Informatics—Computing Technology, Intelligent Technology, Industrial Information, Integration. pp. 48–51. IEEE (2016)
9. Dhanapal, A., Saravanakumar M.V., Sabibullah. M.: Emerging big data storage architectures: a new paradigm. i-Manag. J. Pattern Recogn. **4**(2), 31–41 (2017)
10. Raja, K., Sabibullah, M.: Big data driven cloud security—a survey. In: IOP Conference Series, Materials Science & Engineering (ICMAEM-2017), vol. 225 (2017)
11. Arulananthan, C., Sabibullah, M.: Smart Health- Potential & Pathways -A Survey, Vol. 225, IOP Conference Series, Materials Science & Engineering (ICMAEM-2017) (2017)
12. Ghaffar, A., et al.: Big data analysis: an spark perspective. Glob. J. Comput. Sci. Technol. Softw. Data Eng. Version 1.0 **15**(1) (2015)

Recent Developments in Big Data Analysis Tools and Apache Spark

Subhash Chandra Pandey

Abstract With the explosion of data sizes, the domain of big data (BD) is gaining enormous and prevalent popularity and research worldwide. The big data as well as big data repository possesses some peculiar attributes. Perhaps, analysis of big data is a common phenomenon in today's scenario and there are many approaches with positive aspects for this purpose. However, they lack the support to deal conceptual level. There are numerous challenges related to the performance of BD analysis. Precisely, these challenges are mainly related to enhance the effectiveness of big data analysis and optimum utilization of resources. Indeed, the lack of runtime level indicates the unawareness of various uncertainties pertaining to analysis. Furthermore, maneuvering analytic uncertainty is a challenging task and it can be created because of many reasons such as due to the fusion of huge amount of data from different sources. In addition, analytic uncertainty is also hard to predict the aspects for which the data is useful for the purpose of analysis. The main focus of this chapter is to illustrate different tools used for the analysis of big data in general and Apache Spark (AS) in particular. The data structure used in AS is Spark-RDD and it also uses Hadoop. This chapter also entails merits, demerits, and different components of AS tool.

Keywords Big data · Big data analytical tool · Analytic uncertainty
Apache Sparks (AS)

1 Introduction

In modern perspective humans life is substantially depend on technical devices. Further, different tasks of our day-to-day life can be completed easily by the use of technology. However, interaction with technical devices emerging forth the need to deal with data. Indeed, data can be considered as the innate resource. Perhaps,

S. C. Pandey (✉)
Birla Institute of Technology, Mesra, Ranchi (Patna Campus), Bihar, India
e-mail: Subh63@yahoo.co.in

© Springer Nature Singapore Pte Ltd. 2019
M. Mittal et al. (eds.), *Big Data Processing Using Spark in Cloud*,
Studies in Big Data 43, https://doi.org/10.1007/978-981-13-0550-4_10

imbibing and performing different sorts of processing on the data creates many important and innovative aspects of the objects. It also render valuable insight pertaining to world [1]. Furthermore, the domain of data management is observing exponential growth, speed, and variety in the data and thus the data management is facing new challenges. Indeed, it is beyond imagination to perform the study of a complex system without the use of data in any scientific endeavor. In addition, data science is also important in different walks of life as well as in industries. Perhaps, big data accessing, big data analysis, and many other issues related with big data creates substantial technical challenges before the researchers. Very often in real practice we gather the data from different sources and then manage then in meaning full way. Indeed, the domain of big data incorporates different aspects such as accumulation of data from various domains as well as revamping them in effective ways. Further, big data analysis also incompasses the task of discovering the useful information from the accumulated data. In spite of exposing the hidden facts and figures, the big data analysis can also displays the categorization of the data on the basis of knowledge extracted. Precisely, it is called knowledge discovery from the big data.

The entire scenario of big data has drastically been changed after the emergence of cloud computing. The big data and its hybridization with cloud computing renders the sharing and consumption of huge quantum of data from a diversity of dispersed resources. Perhaps, BD analytics ecosystem is vital aspect of this hybridization. The BD analytics ecosystem can be considered as cluster of methods. Further, BD analytics ecosystem is also treated as middleware. It is mentioned in [2, 3] that BD analytic ecosystem is peculiarly designed for understanding the unstructured data. Perhaps, the target of BD analysis is to work out more succinct information from the BD. This domain of computation has created many new aspects and it eventually triggered a chain reaction [4]. However, collecting more data is advantageous but it is always associated with some short comings such as from user's point of view it will be a hard task to know in advance the usefulness of the data. It is also difficult to predict in advance whether entire of the data or its part is useful. It would also be difficult to visualize in advance that which algorithm work best. In literature this phenomenon is termed as "analytic uncertainty". Further, due to the "analytic uncertainty" it is often possible to entrapped in an ordered trial-and-error process such as application of data transformation, interpretation of result, algorithmic adjustment, and repeat. Author would like to present the example of Twitter. Twitter as an organization processes massive amount of data. The average data processing rate of twitter is considerably high. Twitter processed entire data with substantially fast pace and they also attempt to reduce the 'analytic uncertainty' so that different ethical policies can be maintained. However, if we contemplate the entire data market explosion, we can observe the fact that reduction in 'analytic uncertainty' is still occurring with a slow pace and is really a difficult task to yield requisite outputs in a fashion coherent with time. It is pertinent to mention that majority of organizations are interested in Hadoop for the purpose of data analysis because Hadoop possesses some outstanding attributes. These attributes are: First, it is an easily programmable model. This programming model is known as MapReduce. Second, it incorporates many features such as:

- Scalability
- Flexibility
- More tolerant to fault
- Reduced cost

The important factors in the processing of big datasets are (1) time interval of waiting between successive queries and (2) waiting time required for executing the program. In this chapter, author will present many pragmatic issues related to analysis of BD. Different important features of this chapter are; brief introductory notes on innovative BD analytics ecosystem. This chapter also presents miscellaneous aspects of uncertainty relate to analysis.

The main focus of this chapter is to discuss the different perspective of tool used for the analysis of big data. This chapter is mainly concerned with AS (Apache Spark) tool. This tool uses Hadoop. Indeed, AS is based on an advanced computing technology known as cluster computing. AS tool has some important features. These features are described in detail in Sect. 5 of this chapter. Indeed, Spark is neither an advancement and updation nor dependent of Hadoop. Furthermore, Spark supports in-memory computing. The in-memory computing enables Spark processing swift in comparison other disk based techniques such as Hadoop. Some other advantages of Spark are mentioned in [5]. The use of Hadoop in Spark is two fold. These are (1) storage and (2) processing. It is worthy to mention that Spark uses Hadoop only for storage purpose.

This chapter organized as follows: Sect. 2 focuses on big data and its repository. Section 3 describes big data analytics approaches and uncertainty. Further, Sect. 4 renders a comparative study of big data analytics tool. A lucid discussion of AS is given in Sect. 5. Moreover, a fundamental and peculiar data structure used in AS is described in Sect. 6. Finally, chapter is concluded in Sect. 7.

2 BD and Its Repository

The recent trend of data mining is the analysis of big data. Big data differs from ordinary datasets in many respects. Many important features of big data is given in Sect. 2.1. However, exponential growth and all possible variation in the huge quantity of the data are the essential features of big data. Further, the topic of big data analysis is treated as a globally prevalent topic of interest. It prevails and equally important for its efficacy in the business domain and society as the Internet has been. In general, it is an well accepted and proven fact that accurate analysis can be obtained with the help of more data. Further, more accurate analysis could produce an outcome which will be more legitimate, timely and could help in confident decision-making. Thus, big data helps to maintain the higher operational efficiencies with reduction in risk and cost [6].

2.1 BD

BD incorporates many salient features. Some important features of big data are enumerated as follows:

a. Volume-wise
b. Velocity-wise
c. Variety-wise
d. Variability-wise
e. Complexity wise.

2.1.1 Volume-Wise

Volume is core and relevant issue of the big data. In many sectors, it is necessary to store the huge quantity of big data for decades. Perhaps, the archival of huge volume of big data for decades is difficult, cumbersome, and costly affair. Indeed, data storage is a big issue. Further, this issue has almost been resolved by the advance and affordable storage devices. Further, due to the great deal of research in hardware domain happened in recent past big data storage itself is not a big challenge. However, in contrast, huge volume are still emerging forth some other challenges. One such challenge can be to maintain the compatibility between the huge data volume and valuability of the information stored [7].

2.1.2 Velocity-Wise

Indeed, many challenges of varying nature are associated with the issue of big data. Indeed, besides volume velocity is also an important factor pertaining to big data. Further, the swift augmentation in the volume of big data is also creating challenges related to time and efficiency. Modern technologies such as Internet streaming, RFID tags, automation and sensors, robotics, etc., are still persisting to cope with this increasing in the velocity of the big data. Perhaps, it is an essential requirement to deal with the data. It is mentioned in [8] that velocity is also a major challenge in the realm of big data.

2.1.3 Variety-Wise

Managing the high variance in big data is also a challenging task. In fact, it can be considered as a bigger challenge than the volume and velocity. As a matter of fact, enormous growth in data can take place in variety of ways such as structured, unstructured, etc. Further, it is really a challenging task to work out the effective means to establish the inter-relation among the discrete data in time so that required information from the data can be extracted. Indeed, it is a strenuous task and huge numbers of public and private bodies are trying hard for obtaining a comparatively good and effective means for coping with this issue [7].

2.1.4 Variability-Wise

Undoubtedly, the rapid growth and consistent increase in variety render dealing with big data challenging. However, frequent growth and decay is also a substantial challenge. The example of this situation can be envision as the responses on social media related to global events. Responses on social media create huge volume of data and the analysis of this huge data must be accomplished before the pattern altered. Furthermore, different phenomena also impinge thrust on fiscal domain and it causes considerable effect on unstructured data [9]. Therefore, it is obvious that variability plays vital role while dealing with big data.

2.1.5 Complexity-Wise

Different issues discussed above turn the big data a challenging entity. The complexity of big data due to huge volume synergized by increased variety of sources and with unpredicted trends make the data mining tasks hard with big data. However, in spite of all the constraints, different analytic tools have been developed to analyze the big data. Attempts have already been made to develop the techniques for establishing the relational pedigree and connectivity so that pertinent information can be extracted before the data become uncontrolled. This is required so that catastrophic losses can be avoided. The complexity involved in big data is an important issue in today's scenario [9].

2.2 Characteristics of Big Data Repository

Precisely, in order to declare a data repository as a big data it must possess certain characteristics. Big data must be accessible with satisfactory possibility of uses. Further, big data must be extensible. In addition, big data contains petabyte or more data. From extensive literature survey, it has been observed that big data must possess the characteristics as enumerated below [10].

- Accessible
- No distributed management
- No centralized unuseful data storage
- Possibilities of future extension
- Extremely fast data insertion
- Handles large amounts
- Reliable hardware

- Not very costly
- Enable to process the data parallely
- Fast data processing.

3 Big Data Analytics Approaches and Uncertainty

In this section, author will present different approaches pertaining to the analysis of BD. This section is divided into two sub-sections, i.e., Sects. 3.1 and 3.2. Section 3.1 describes the big data analytic approaches and subsequently Sect. 3.2 will present different emerging trends of analytic uncertainty of BD.

3.1 Analytic Approaches

Mainly, there are three analytic approaches for the analysis of big data. These are

- MapReduce-Like
- In-Memory Generic Processing (IMGP)
- Stream Processing.

3.1.1 MapReduce-Like

MapReduce model (MRM) [11] paradigm adapts a peculiar design approach. This design approach incorporates automated parallelization which in turn renders its use easy. This approach is often termed as the data-locality centered design approach. In this approach simulation can be programmed in accordance with to the data. This is because of the fact that elements of computing and the location of data is not hidden. Moreover, in this paradigm the storage layer and various elements of computing are co-located. This approach provides a substantial scalable system and is easy to use.

3.1.2 IMGP

As is already mentioned, this approach is easy to use. However, many limitations related to expressivity and performance have been observed with mapReduce. Thus, a new approach for data analysis is emerged which is known as 'in-memory big data analytics' approach. In this approach, considerable use of IMGP takes place. The reduction in interactions subsequently decreases the bottlenecking of inputs and outputs. This is because of the following reasons:

- Slow local disks
- Extra copies
- Serialization issues

In order to enhance the meaningfulness, in this approach, multi-step data pipe lines are facilitated. This task is accomplished by employing directed acyclic graph (DAG). The DAG works as a runtime construct. It also uses high-level control flow of the application. One illustrative example of this domain is Spark [12].

3.1.3   Stream Processing

Many data sources are live. Live data sources play important role in big data analytics. Indeed, live data sources incorporates many important characteristics such as online data processing. Therefore, live data produces enhanced reactivity and cleanness of the final outputs. Further, it eventually leads a better insight about the data processing. Perhaps, this is the main reason behind the rapid growth of big data processing. In [3], such an engine is described. This engine presents basic implications in lucid terms. Further, some highly reactive and sophisticated approaches for processing have been grew keeping in view the requirements of the time. These approaches are mainly S4 [13], MillWheel [14], ApacheFlink. These approaches display different issues. However, these approaches mainly addresses issues as given below.

- Fault tolerance
- Low latency
- Consistency

It is pertinent to mention that low latency is an important factor. In [15, 16], it is mentioned that this trend was propagated by many organizations. Moreover, much emphasis has been given by the researchers so that the approach can include revivness of input data as well as consequences by designing the window concept. Indeed, windows are stream operators. Further, window renders a means to identify that at which point of time ancient data is considered as redundant. This is known as eviction policy. Stream operators also provides mechanism to decide that at which point of time the computation commenced on accumulated data. Moreover, some provision is also given in stream operator to decide the function necessary for computation. Furthermore, stream operator also manages intermediate states that are responsible for producing the final result. This process also incorporates the means required for the duplication of data. It is given in [17–19] that different wired applications are based on the hypothesis of window. Extensive literature survey revealed the fact that substantial research has been performed at the level of performance optimizations [17] expressivity [20].

3.2 BD Use Cases and Analytics Uncertainty

Indeed, BD analytics process is treated as a data-oriented process. It is worthy to mention that data analytics process requires perpetual adaptation evolution. This characteristics differentiate it with the traditional approach. In traditional approaches the data is required to adapt the application. The three illustrative use case considered in this section are given below. Perhaps, in these three use case the analytics uncertainty plays vital role.

- A/B Testing
- Machine Learning (ML)
- Deep Exploration (DE).

3.2.1 A/B Testing

Very often, within the purview of BD it is difficult to tackle different problems while running in production. The trade-off is between the delivery of consistent results and it is required to have continuous refinement. This renders the maximization of delivered value. Here constantly refinement implicates that further value can be extracted. Perhaps, considerable developments in this methods is required so that high rate can take place. However, these ways are insufficient for BD analysis. This is the main reason behind the use of this testing.

3.2.2 ML

Machine learning is domain of paramount importance and in this domain the application of big data analytics substantially take place. It is pertinent to mention that machine learning also uses training phase for the model similar to A/B testing. This stable model is sometime termed as the champion model. Further, machine learning also strives to persistently improve it. Moreover, in order to improve performance of this model an alternative model is constructed. This alternative model is trained with different data. The alternative model can also be improved by incorporating diversification in chief algorithm. Sometimes a hybridization of different approaches to improve the performance of this alternative model is used. The performance of the alternative model or challenger model thus modified can challenge the champion model. There are many challenges in this task. However, execution of both model in a faultless way is the main challenge.

3.2.3 DE

BD can be considered a resilient mean for discovering the patterns and insight from the data. There is a subtle dichotomy between big data and machine learning.

This subtle differentiation is based on the goal of the analysis. The synthesis of data implicates conversion of data into information followed by knowledge. The process of data synthesis entails a range of premise as well as investigation processes over the data [21].

3.2.4 Analytics Uncertainty

The approaches described in Sect. 3.1 have many positive aspects. However, these approaches have certain draw backs at conceptual level. Further, these approaches also lack in runtime level. The lack of runtime level indicates different shortcomings such as the unawareness of aim of the purposes. Therefore, runtime level lacks the optimization opportunities. Following two uncertainties can be considered in this context:

- The breadth uncertainty
- The temporal uncertainty

Often, simultaneous processing of same dataset with multiple algorithms is required. However, in such scenario users have to deal with multifaceted work flows as well as midway states. There are many pertinent issues in such situation like:

- The point of time to divert into an other direction.
- To recognize the intermediate states for the purpose of comparison.
- To decide where to turn back to have a different direction, etc.

However, due to lack of runtime support users are bound to act together with miscellaneous layers clearly. Further, lacking support from the runtime can lead different issues related to scalability and performance. Moreover, due to the unawareness regarding the inter-relationship among different directions runtime occasionally renders non-optimal performance. The obvious example of this is unnecessary duplication of the data. Indeed, the optimization of runtime can be considered as an interesting pursuit of future research.

Further, there is one another aspect of analytics uncertainty related with the temporal dimension. This analytic uncertainty is termed as temporal uncertainty. Precisely, the accumulation of new data is associated with the issue of liveliness It is inquisitive to think that is it possible to process entire of available data? Furthermore, it is also inquisitive to think that is it worthy to process merely the recent data to obtain the substantial useful insight of the future? Moreover, it is also of the paramount importance to ponder that where is the right trade-off. Indeed, it is suggestive to use discrete runtimes for discrete tasks. In fact, it becomes a concern of considerable interest. In addition, users have to use different runtimes for different jobs become an issue of considerable interest due to the fact that the runtimes do not talk to each other and as thus perform sub-optimally. The example of non-communicative runtimes is sharing common data and states.

4 Different Analytical Tools

In this section, author would like the introduce some frequently used analytic tools within the purview of big data. Each tools have their own merits and demerits. In addition, this section will also present some substantial information regarding to AS. Moreover, a comparative study of As with other tools will also be incorporated.

- AH (Apache Hive)
- AP (Apache Pig)
- AZ (Apache Zebra)
- AHB (Apache HBase)
- AC (Apache Chukwa)
- ASt (Apache Storm)
- AS (Apache Spark).

4.1 AH

AH is used as an infrastructure for data warehousing and it requires Hadoop for its execution. Apache hive imparts Hive Query Language (HQL). This language organizes, aggregates and enable the processing of queries for data. HQL is analogous to SQL in many aspects [22]. It is different from the Pig Latin language, which uses more procedural approach. There are some common features in SQL and HQL such as:

- Final results are described in big query. In contrast, in PL (Pig Latin) uses a step by step procedure.
- HQL is similar to standard query language and it enables the developer to write the query in HQL. It is convenient for the SQL developers.
- It is possible to break the HQL queries using Hive.
- The breaking of HQL queries enable the communication with MapReduce and thus jobs are executed across a Hadoop Cluster.

4.2 AP

It can be considered as a platform for analyzing the substantial quantity of big data. There are many important feature of AP including parallel processing of the tasks. This key feature fortify the Pig tool to tackle the gigantic datasets [22]. The basic tasks performed by Pig and Hive are same [23]. However, in Pig preparation of the data takes place in an effective manner. And it is also suitable for the following purposes:

- Data warehousing
- Data presentation

In the Pig tool, the sequence of operations are as: (1) collection (2) noise removal with the help of specific Pif tools (3) data is stored. It is also pertinent to mention that in contrast to AP Hive can be use for the purpose of ad hoc queries.

4.3 AZ

It can be treated as a type of storage layer for data access. It performs a great deal of abstraction. AZ also provides a properly arranged view of the data. Thus, provides comfort for the Pig users. Indeed, AZ is treated as a sub-project of AP. It is mentioned in [24] that AZ gives abstraction between PL and the Hadoop Distributed File System. There is a specific feature in Zebra. Further, in [25] it is given that Java testing framework can be used with AZ.

4.4 AHB

It is a database engine and built with the help of Hadoop. Besides other characteristics, this tool also renders interfacing with Pig and Hive. A storage function of Pig API is used in AHB. In [26], reason behind this phenomenon is given. It is due to the fact that the chunk of data are substantially small and thus there is no need to use of HBase.

4.5 AC

It is a monitoring system. It is mainly used for processing the log files which are obtained from especially from Hadoop and other distributed systems [26]. The diversity of Chukwa tool is less than the Pig tool. Indeed, it is designed for very confined area of data processing and its scope is not considerably wide.

4.6 ASt

It uses distributed computing. Further, researcher are planning to use this tool with Apache. Further, it can be considered as an effective tool to process gigantic stream of knowledge from big data. Furthermore, it was primitively used to process the twitter streams. Moreover, it has many attracting features. These are

- It is reliable
- It is scalable
- Assured information transformation
- Deployment and utilization is easy [5].

The crux of this chapter is to discuss AS for the task of big data analysis. Therefore, author would like to give sufficient impetus to AS and thus allocate a separate section, i.e., Sect. 5 for AS.

5 Apache Spark

AS is Hadoop's sub-project. Further, it was presented as an open sourced software in 2010. It also augments the versatility of MRM and thus it renders efficient use of MRM for more types of computations. This extended use of MapReduce also incorporates interactive queries and stream processing. Further, AS possesses many salient features such as

- It has provision of interfacing with many programming languages.
- In-memory cluster computing is the main characteristic of Spark.
- AS is considerably swifter in comparison to traditional techniques due to in- memory processing.
- The applicability of AS is versatile and it can be used for a wide range of workloads.
- AS can be efficiently used for many applications.
- AS also reduces management burden of maintaining different tools for different workload.
- Its functioning is different from other tools such as Hadoop.

5.1 Why Apache Spark?

Apache Spark is a versatile tool and it entails many distinguished features which make it enable to differ from other available tools. Some important attributes of Apache Spark which differentiate it from other tools are given below.

- It is a speedy system [5].
- AS uses parallel data processing system.
- Workflows are defined in an evocative style as in Map Reduce.
- High capability in comparison to other tools.
- It can easily be installed.
- It can implemented in Scala [25].
- Spark streaming is supported by many organizations.
- It is supported by many large companies.

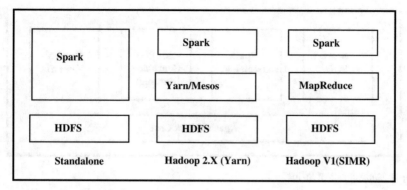

Fig. 1 Spark built with Hadoop component

- The reliability of Spark system used in healthcare solutions can be judged from Intel recommendation [27].
- It is supported and funded by many open-source contributors [5].
- Apache Spark is considered as Data Science tool [26].
- It can connect datasets across numerous dissimilar data avenues.

5.2 Draw Backs of AS

The working of AS is substantially fast. It is also considerably versatile and can be used for the wide range of big data analysis. However, it is hard to consider any system 100% perfect and flawless. Some imprecision and pitfalls will always be there. Likewise, there are two constraints in AS also which preclude its suitability [28]. These constraints are given below.

- Low Tolerance to Latency requirements
- Shortage of Memory resources.

5.3 Spark Built on Hadoop

Spark is Hadoop's sub-project and three approaches can be used to explain it. These are

- The approach of Standalone
- The approach of Hadoop Yarn
- Spark approach in MapReduce (SIMR)

These three ways of Spark development is shown in Fig. 1.

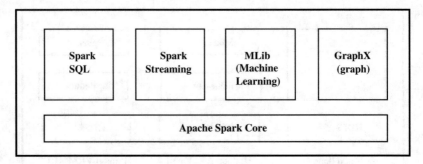

Fig. 2 Components of Spark

5.4 Components of Spark

The components of Apache Spark (s) are given below.

- Apache Spark Core (APC)
- S-SQL
- SS (S-Streaming)
- MLlib (Machine Learning Library)
- GraphX.

APC is the corer component for spark platform. Further, different other components are structured over the APC. Further, it displays the characteristics of in-memory computing. Moreover, it uses external storage systems. Further, the component SSQL is situated upon the APC and it preludes SchemaRDD. The SchemaRDD renders an innovative abstraction for the data. And both structured as well as unstructured data is supported by it. Furthermore, the third component, i.e., Spark Streaming increases the fast scheduling capability of Spark Core so that streaming analytics can be performed. A detailed discussion about RDD is given in Sect. 6. Moreover, MLlib is the fourth component. Further, GraphX is the fifth component. The functioning of GraphX is also to provide an API. Different components of Spark are shown in Fig. 2.

6 Spark-RDD

In Spark a fundamental and peculiar data structure is used. This data structure is commonly termed as Resilient Distributed Datasets (RDD). Further, it holds provenance information which are called as lineage. Indeed, there are two approaches for its generation. In addition, it can be reconstructed by applying the partial recomputation approach of its ancestors. There can be variety of external storage system for referencing a data like shared file system, HDFS, HBase, etc. It is also important to mention that data source presenting HIP (Hadoop Input Format) can also work as an external storage system in this context. RDD operations can be classified into two categories. These are given below.

- Narrow: In such RDD operations, the output partition is simulated directly from their input partition.
- Wide: In such RDD operations, the output partition is simulated just by assembling different input patterns.

Pragmatically, wide transformation is considered to be more challenging. This is due to the fact that this transformation entails crucial issues. Further, this complex data exchange may render different characteristics. In [29], different characteristics are enumerated such as performance, scalability, and resource utilization. In addition, certain other issues have also been discussed in [30]. There are some other important associated features of RDD.

Indeed, Spark uses the concept of RDD and the aim behind this is to achieve faster and efficient MapReduce operations. MapReduce is an efficient and prevalent tool. Some important characteristics of MapReduce are as under.

- MapReduces is used for the data processing and for generating the large datasets as well.
- MapReduce uses parallel, distributed algorithm on a cluster for this purpose.
- MapReduce permits the parallel computations. Moreover, it does so without incurring any sort of worriness about work distribution and fault tolerance.
- MapReduce permits various abstractions.
- MapReduce permits both iterative as well as interactive implementations.
- In it data sharing is slow.

It is imperative and pertinent to mention that reuse of the results is possible in MapReduce. The reuse of the result can pragmatically takes place across different implementations. Working of iterative operations in MapReduce are illustrated in Fig. 3. Further, Fig. 4 explains the working of interactive queries on MRM.

6.1   Data Sharing Using Spark-RDD

The process of replication, serialization, and disk I/O renders the data sharing operation slow in MapReduce. Further, in order to rectify this drawback researchers developed a specialized framework. This peculiar framework is called AS. The central functioning of spark is based upon the 'resilient distributed datasets (RDD)'. The functioning of spark supports in—memory processing. It is innocuous and in-memory renders substantial faster data sharing. Now, author would like to introduce the functioning of iterative and interactive operations in spark-RDD. Figure 5 shows the functioning of iterative operations on Spark-RDD. Storing the intermediate results in a distributed memory rather than on disk renders the system faster. Further, the illustration given in Fig. 6 shows interactive operations on Spark-RDD. It should be

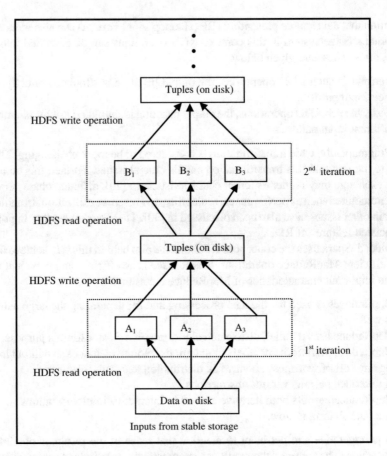

Fig. 3 Flow of iterative operations in MapReduce

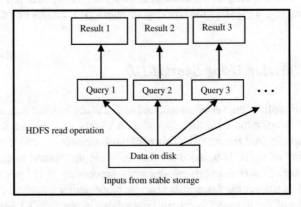

Fig. 4 Flow of interactive operations in MapReduce

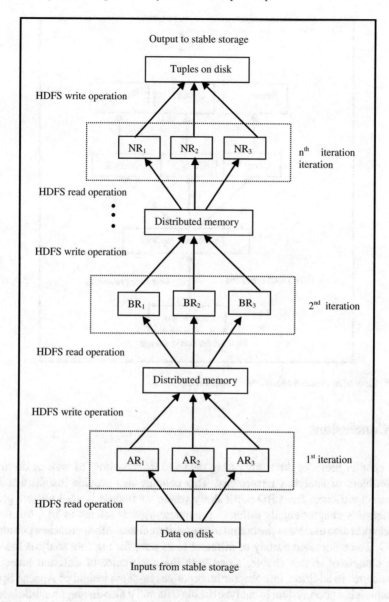

Fig. 5 Flow of iterative operations in Spark-RDD

noted that very often different queries are executed on the same dataset repeatedly. Therefore, it is incisive to keep this particular data in memory to observe a better execution times. Moreover, provision for support to persist RDDs on disk is also provided. There is also support for persisting RDDs replication across multiple nodes.

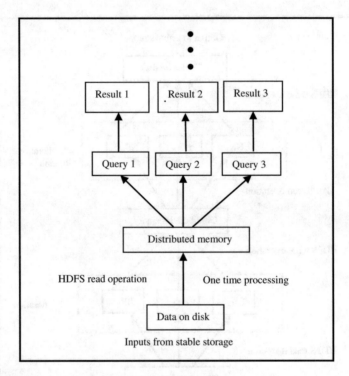

Fig. 6 Flow of interactive operations in Spark-RDD

7 Conclusions

This chapter preludes the study of big data and its repository as well as discusses the problem of analytics uncertainty. This chapter also provide insight that how to extract attributes from BD particularly where usefulness is not known a priori. Further, this chapter entails different pragmatic aspects related to BD. Moreover, this chapter also discusses the boundaries of different analytic approaches pertaining to BD. The comparative study of different tools used for big data analysis has also been elucidated in this chapter along with the importance of different indecisive dimensions. In addition, this chapter incorporates in-depth insight of Apache Spark. Perhaps, it is a good system to analyze the big data and it also displays a subtle latency of few seconds. Furthermore, this chapter also vividly preludes resilient distributed datasets. This a specific type of fundamental data structure used in Spark. Moreover, some attributes of this chapter can be considered as the future research direction in this pursuit. Two features in the realm of big data analysis are acknowledged as indispensable future research direction. Perhaps, the first feature is regarding the working with AS. It is not easy to work with AS. Therefore, some substantial research must be perform so that this drawback can be rectified.

References

1. Hey, T., Tansley, S., Tolle, K.M.: The Fourth Paradigm: Data-Intensive ScientificDiscovery. Microsoft Research, Redmond (2009)
2. Flink. https://flink.apache.org/
3. Toshniwal, A., et al.: Storm@twitter. In: 2014 ACM SIGMOD International Conference on Management of Data, SIGMOD 2014, pp. 147–156. Snowbird, USA (2014)
4. The Zettabyte Era: Trends and Analysis. Cisco Systems, White Paper 1465272001812119 (2016)
5. Community effort driving standardization of Apache Spark through expanded role in Hadoop Project, Cloudera, Databricks, IBM, Intel, and Map R, Open Source Standards. http://finance.yahoo.com/news/communityeffortdrivingstandardizationapache162000526.html. Accessed 1 July 2014
6. Big Data: what I is and why it mater (2014). http://www.sas.com/en_us/insights/big-data/whatis-big-data.html
7. Lewis, N.: Information security threat questions (2014)
8. Goldberg, M.: Cloud Security Alliance Lists 10 Big data security Challenges (2012) http://data-informed.com/cloud-security-alliance-lists-10-big-data-security-challenges/
9. Securosis, Securing Big Data: Security Recommendations for Hadoop and No SQL Environment (2012). https://securosis.com/assets/library/reports/SecuringBigData_FINAL.pdf
10. Hurst, S.: To 10 Security Challenges for 2013 (2013). http://www.scmagazine.com/top-10-security-challenges-for-2013/article/281519/
11. Dean, J., Ghemawat, S.: Mapreduce: Simplified data processing on large clusters. In: 6th Conference on Symposium on Operating Systems Design and Implementation (OSDI), pp. 1–13. USENIX Association, San Francisco (2004)
12. Zaharia, M., Chowdhury, M., Das, T., Dave, A., Ma, J., McCauly, M., Franklin, M.J., Shenker, S., Stoica, I.: Resilient distributed datasets: A fault-tolerant abstraction for in-memory cluster computing. In: The 9th USENIX Symposium on Networked Systems Design and Implementation, NSDI 2012, San Jose, USA (2012)
13. Neumeyer, L., Robbins, B., Kesari, A., Nair, A.: S4: Distributed stream computing platform. In: 10th IEEE International Conference on Data Mining Workshops, ICDMW 2010, pp. 170–177. Los Alamitos, USA (2010)
14. Akidau, T., Balikov, A., Bekiroglu, K., Chernyak, S., Haberman, J., Lax, R., McVeety, S., Mills, D., Nordstrom, P., Whittle, S.: Millwheel: fault-tolerant stream processing at internet scale. In: Very Large Data Bases, pp. 734–746 (2013)
15. Zaharia, M., Das, T., Li, H., Shenker, S., Stoica, I.: Discretized streams: An efficient and fault-tolerant model for stream processing on large clusters. In: 4th USENIX Conference on Hot Topics in Cloud Computing, HotCloud 2012 (2012)
16. Tudoran, R., Costan, A., Nano, O., Santos, I., Soncu, H., Antoniu, G.: Jetstream: enabling high throughput live event streaming on multi-site clouds. Future Gener. Comput. Syst. **54**, 274–291 (2016)
17. Carbone, P., Traub, J., Katsifodimos, A., Haridi, S., Markl, V.: Cutty: aggregate sharing for user-defined windows. In: 25th ACM International on Conference on Information and Knowledge Management, CIKM, pp. 1201–1210 (2016)
18. Hammad, M.A., Aref, W.G., Elmagarmid, A.K.: Query processing of multi-way stream window joins. VLDB J. **17**(3), 469–488 (2008)
19. Yang, D., Rundensteiner, E.A., Ward, M.O.: Shared execution strategy for neighbor-based pattern mining requests over streaming windows. ACM Trans. Database Syst. **37**(1), 5:1–44 (2012)
20. Akidau, T., Bradshaw, R., Chambers, C., Chernyak, S., Fernndez-Moctezuma, R.J., Lax, R., McVeety, S., Mills, D., Perry, F., Schmidt, E., Whittle, S.: The data flow model: a practical approach to balancing correctness, latency, andcostin massive-scale, unbounded, out-of-order data processing. Proc. VLDB Endow. **8**, 1792–1803 (2015)

21. Cao, L., Wei, M., Yang, D., Rundensteiner, E.A.: Online outlier exploration over large datasets. In: 21th ACM SIGKDD International Conference on Knowledge Discovery and Data Mining, KDD 2015, Sydney, Australia, pp. 89–98 (2015)

22. Hoover, M.: Do you know big data's top 9 challenges? (2013). http://washingtontechnology.com/articles/2013/02/28/big-datachallenges.aspx

23. MarketWired (2014). http://www.marketwired.com/press-release/apache-spark-beats-the-world-record-forfastest-processing-of-big-data-1956518.htm

24. Donkin, R.B., Hadoop And Friends (2014). http://people.apache.org/~rdonkin/hadooptalk/hadoop.html. Accessed May 2014

25. Hadoop, Welcome to Apache Hadoop, (2014). http://hadoop.apache.org/. Accessed May 2014

26. Stella, C.: Spark for Data Science: A Case Study (2014). http://hortonworks.com/blog/spark-data-science-case-study/

27. Basu, A.: Real-Time Healthcare Analytics on Apache Hadoop using Spark and Shark (2014). http://www.intel.com/content/dam/www/public/uen/documents/white-papers/big-data-realtimehealthcare-analyticswhitepaper.pdf. Accessed Dec 2014

28. Spark MLib: Apache Spark performance (2014). https://spark.apache.org/mllib/. Accessed Oct 2014

29. Nicolae, B., Costa, C., Misale, C., Katrinis, K., Park, Y.: Leveraging adaptive I/O to optimize collective data shuffling patterns for big data analytics. IEEE Trans. Parallel Distrib., Syst (2017)

30. Nicolae, B., Kochut, A., Karve, A.: Towards scalable on-demand collective data access in IaaS clouds: An adaptive collaborative content exchange proposal. J. Parallel Distrib. Comput. **87**, 67–79 (2016)

SCSI: Real-Time Data Analysis with Cassandra and Spark

Archana A. Chaudhari and Preeti Mulay

Highlights

- The open-source framework for stream processing and enormous information
- In-memory handling model executed with the machine learning algorithms
- The data used in subset of non-distributed mode is better than using all data in distributed mode
- The Apache Spark platform handles big data sets with immaculate parallel speedup.

Abstract The dynamic progress in the nature of pervasive computing datasets has been main motivation for development of the NoSQL model. The devices having capability of executing "Internet of Things" (IoT) concepts are producing massive amount of data in various forms (structured and unstructured). To handle this IoT data with traditional database schemes is impracticable and expensive. The large-scale unstructured data required as the prerequisites for a preparing pipeline, which flawlessly consolidating the NoSQL storage model such as Apache Cassandra and a Big Data processing platform such as Apache Spark. The Apache Spark is the data-intensive computing paradigm, which allows users to write the applications in various high-level programming languages including Java, Scala, R, Python, etc. The Spark Streaming module receives live input data streams and divides that data into batches by using the Map and Reduce operations. This research presents a novel and scalable approaches called "*Smart Cassandra Spark Integration (SCSI)*" for solving the challenge of integrating NoSQL data stores like Apache Cassandra with Apache Spark to manage distributed systems based on varied platter of amalgamation of current technologies, IT enabled devices, etc., while eliminating complexity and risk. In this chapter, for performance evaluations, SCSI Streaming framework is compared

A. A. Chaudhari
Symbiosis International (Deemed University), Pune, India
e-mail: chaudhari.archana12@gmail.com

P. Mulay (✉)
Department of CS, Symbiosis Institute of Technology, Symbiosis International (Deemed University), Pune, India
e-mail: preeti.mulay@sitpune.edu.in

© Springer Nature Singapore Pte Ltd. 2019
M. Mittal et al. (eds.), *Big Data Processing Using Spark in Cloud*,
Studies in Big Data 43, https://doi.org/10.1007/978-981-13-0550-4_11

with the *file system-based data stores* such as Hadoop Streaming framework. SCSI framework proved scalable, efficient, and accurate while computing big streams of IoT data.

Keywords Apache Spark · Apache Cassandra · Big data · IoT · MapReduce

1 Introduction

Best way to predict the future is to create it!

—Peter F. Drucker

As mention in above proverb, the writer advises that with the help of technologies it is best to create future in addition to predictions based on Data Mining, Data Warehouse and other related techniques. This is also true with Incremental Learning, *Internet of Everything (IoE)*, and the new innovation and methodologies that have arrived at our doorstep.

Internet of Things (IoT) [1] is very widely used concept and computations around the world. The administration, the scholarly community, and industry are engaged with various parts of research, execution, and business with IoT. The domains in which IoT-based systems are widely used include agriculture, space, health care, manufacturing, construction, water, and mining, etc., to name a few. IoT-based applications such as digitizing agriculture, electric power and human interaction, smart shopping system, electric foundation management in both urban and provincial regions, remote health care monitoring and notification systems, and smart transportation systems are widely used and becoming popular.

The development of data created by means of IoT has assumed a noteworthy part in the big data landscape. The prevalence of IoT makes big data analytics challenging because of handling and accumulation of information from various sensors in the IoT environment. In IoT, different sensors and multivariate objects are involved in data generation, which includes variety, noise, abnormality, and quick development. The data generated using IoT-based devices have different formats as compared to the big data. The list of new dimensions is added to the big data recently and they are popularly known as "7 Vs of Big Data". In this chapter, the Electricity Smart Meter data is considered for research study. This Electricity Smart Meter is based on IoT, GPS, and Cloud Computing concepts. The more elaborate discussion is given in following sections related to Electricity Smart Meter. Based on above discussion, and research it is observed that the same "7 Vs of Big Data" can be mapped directly to multidimensional IoT Data, in which seven estimations relate to the basic parts of IoT/IoE data [2] as shown in Fig. 1.

The dimensions of IoT data can be defined as follows:

1. Volume: The IoT Data comes in an expansive size that is Petabytes, Gigabytes, Zettabytes (ZB) and Yottabytes (YB). The IoT makes exponential development in data.

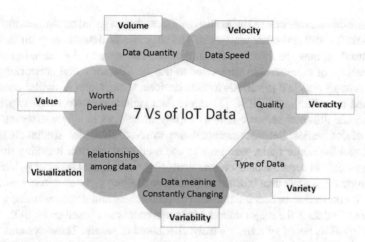

Fig. 1 The proposed dimensions of Internet of Thing (IoT) Data

2. Velocity: This is the most important thing related to dynamic data creation and its speed, along with data which is created, delivered, made, or invigorated.

3. Veracity: The huge amount of IoT data collected for data analysis purposes can lead to factual mistakes and distortion of the collected information. Veracity alludes more to the provenance or unwavering quality of the information source, its context, and how meaningful it is to the analysis based on it.

4. Variety: Big Data reaches out past organized information to incorporate unstructured information of all assortments, for example, as text, audio, video, log files, and more.

5. Variability: Variability is different from variety. Variability means if the significance of the information is continually transforming it can affect your information homogenization.

6. Visualization: Visualization refers to the relationship among data. Utilizing outlines and charts to imagine a lot of complex information is substantially much more effective in conveying meaning than spreadsheets and reports crammed with numbers and equations.

7. Value: Value is the capacity to change over IoT data into a concerning reward.

These 7Vs of IoT data layout is the direction to analytics, with each having inborn value in the process of discovering researchful details. The value of these IoT data is considered only when some useful information is extracted from it which is further used in decision making and pattern recognition, etc., by experts and analysts.

The dynamic and huge progress related to different categories of sensors related to IoT, scientific and industrial datasets are the main impetus behind the development of NoSQL model. As an example, in the IoT environment processing and collecting the data through different sensors including self-broadcasting sensors too.

A huge datasets are created by exchanging thoughts on social media sites including Facebook, Twitter and other platforms. The structure of datasets may differ drastically based on how the data is written, created, and stored using varied attributes. The structure of datasets also varies due to use of different social networking platforms, sensors attached to various mobile devices, servers, etc. A similar concept is discussed by authors in NERSC [3] where data generating from single experiment may include different sensors. Due to variety of sensors in single experiment the format of data varies. Before conceptualizing the NoSQL model, similar challenges existed and the major query was how to use methodologies for handling different structured data in separate datasets and analyzing each dataset separately. This way of handling large scale data is difficult to achieve the "big picture" without establishing basic connection between datasets. For predictions and decision-making based on collective data is the major concept for implementation based on NoSQL.

The IoT-/IoE-based systems are truly distributed in nature. These systems create large unstructured and semi-structured data which is growing exponentially and dynamically. The disturbed system approach is necessary for analyzing non-uniform data. The basic aim of an Apache Spark [4] as a model of decision is to effectually handle design and system issues related to Big Data. The author quotes in this [4] paper that Apache Spark offers functionalities to handle various processing and storage of unstructured, semi-structured, and any non-uniform data. However this support is not a straightforward based on query related to data. To extract required knowledge, or to achieve knowledge augmentation, the developing datasets not only need to be queried to permit timely information collection and sharing but also need to experience complex batch information analysis operations.

The NoSQL systems offer not only the capability of storing huge information but also queries to be applied on stored data. This data later analyze and mined for knowledge augmentation, NoSQL Data store is beneficial, as NoSQL data stores significantly handles in the micro-batches, there is a requirement for a product pipeline allowing Big Data processing models such as Apache Spark to tap IoT/IoE data as sources of input.

Summary of Introduction Section

The integrated research discussed in this chapter comprises of different domains viz., Big Data Analytics, Internet of Things (IoT), Database Management System, Distributed System, etc. This research presents a processing pipeline permitting Spark programs written in any language (such as Java, Scala, R, Python) to make utilization of the NoSQL storage frameworks. Such a pipeline is valuable in applications that require real-time processing. Also, presents the dimension of IoT data. This chapter develops "Smart Cassandra Spark Integration (SCSI)". SCSI utilize Apache Cassandra as a data storage, which is open source distributed framework, and consolidate it with Apache Spark as a real-time data processing, which is a speedy and distributed cluster computing system.

The research contribution for this chapter is as follows:

- The proposed SCSI handles real-time-dynamic data/time-series data for running real-time application with the integrated Spark and Cassandra frameworks. The performance and evaluation of SCSI dynamic-data handling is based on MapReduce and Hadoop frameworks.

- The Apache Spark reads the information directly from the different data sources such as empirical multivariate datasets, sensors, time-series datasets and the formatted streaming data is stored into Cassandra for handling high fault-tolerance functionalities.
- The comparative analysis of the execution ramifications of processing data directly to the Apache Cassandra servers versus the use of real-time-dynamic data for processing its takes the snapshot of the database.

2 Background and Shortfalls of Available Approach

2.1 The Cassandra Approach

There are two meaning associated with the term "NoSQL". The first one is "Not only SQL" and other one is "not SQL". NoSQL databases support a subset of SQL commands. It has sprinkling advantages around SQL. The first advantage is there is no need of a rigidly describe schema for the insertion of data directly into the NoSQL database. This data can be changed at anytime, anywhere and these databases [5] adapts itself accordingly. The second advantage is the programmed division and smart detection of the usage of dynamic data. The multiple servers do not cause any problem in NoSQL system. As NoSQL is intelligent and high-level system, but in SQL, the programmer should implement the database schema according to the multiple-storing space arrangements considered in his or her design. The third advantage of a NoSQL data store over a SQL system is that the NoSQL achieves high-speed recovery of data by using caching the data in memory; means a NoSQL system stores valuable data in the cache, just like in computer a processor maintain the cache of recently used resources. Therefore, NoSQL reached a goal of high speed for design and execution of databases, NoSQL is essentially useful for today and future processing of data.

The distributed NoSQL database Cassandra [6] is thoughtfully implemented as non-relational and column-oriented. The initial phase of implementation of Cassandra is by Facebook [7] and now is an Apache project based on open source technologies.

Every Customer-Relationship-Management (CRM) system or Vendible-system generates large amount of dynamic/continuous data. The Cassandra is used to store such huge data effectively. Cassandra uses distributed cluster computing approach to store and handle such data. The cluster computing approach provides flexibility and scalability of the system. The data set consist of columns and rows. The combination of each column and row forms a specific cell having data or unique key related to the dataset. Each column can be sorted using some important key-value pair in all related files, which forms a concept called "keyspace". As mentioned by author in

[8] "the keyspace is alike schema of rational database having families of one or more column in the application". The authors of [8] has also given the fundamental traits related to Cassandra including following:

- Effective MapReduce and Hadoop concept handling,
- Application systems based on HDFS, shared file systems,
- Scalability and redundancy support with fault tolerance,
- Adjustable data partitioning, etc.

2.2 Hadoop and Hadoop Streaming

The Apache Hadoop as discussed by authors in [9] supports an open-source MapReduce framework [10]. The Hadoop Distributed File Systems (HDFS) and MapReduce Frameworks are the two fundamental components of Hadoop Ecosystem. HDFS is used for data storage and management and MapReduce are used for execution of the application. The Hadoop's JobTracker, execute on the Name node or master node. The aim of job tracker is resolving job details that mean a number of Mappers and Reducers required for the execution, listen to the progress of job and status of the worker(s). The DataNode and TaskTracker are located at the worker node. The distributed cluster computing environment supports collection and storage of data into the HDFS and the group of data chunks available for all clusters equally. Further these data-splits/chunks are processed by worker nodes. The local data nodes are responsible for local clustered data chunks.

The most popular platforms Hadoop and MapReduce are implemented using Java programming concepts, Mappers and Reducers utilizes Java APIs. The challenges in front of developers include rewriting legacy application systems using Java programming concepts, to be suitable for Hadoop platform detailing. Hadoop streaming as discussed in [11] provides one of the solutions which allows developers to code MapReduce jobs/Mapper Reducer's operations in any programming or scripting languages, even-though Hadoop streaming has a confined model. The Hadoop streaming is used successfully to implement numerous scientific application systems from numerous domains. These applications system includes space-research related system, complex scientific processing, protein succession correlation, tropical storm detection, air waterway recognition, etc. to name a few.

2.3 Spark and Spark Streaming

The Apache Spark was incubated with the unique formulation of various clusters/splits from input data sets related to Vendible-systems/machines which are important parts of cluster computing, called workers. The Map Reduce concepts also motivate the use of or formulation of workers. The splitting of data and using it in parallel mode is feasible due to user-defined MapReduce functions.

The UC Berkeley AMP Lab is the place where the Apache Spark, one of the famous open-source distributed frameworks is cultured [12]. The Apache Spark is like Hadoop but it is based on the in-memory system to improve overall performance of system, processing, functions, and applications. It is a recognized analytics platform that ensures a fast, easy to use, and flexible computing. The Apache Spark handles complex analysis on large data sets for example, Spark process the programs almost hundred times faster than Apache Hadoop via MapReduce in-memory system.

The Apache Spark is based on the Apache Hive codebase. In order to improve system performance, Spark swap out the physical execution engine of Hive. In addition, Spark offers APIs to support a fast application development in various languages including Java, R, Python, and Scala [13]. Spark is able to work with all files storage systems that are supported by Hadoop. Apache Spark data model [14] is based on the Resilient Distributed Dataset (RDD) [15] abstraction. RDDs constitute a read-only collection of objects stored in system memory across multiple machines. Such objects are available without requiring a disk access. Furthermore, if a partition is lost it can be rebuilt. The Spark project consists of multiple components for recovery of fault, interacting with storage systems, scheduling of task, management of memory, etc.

The Spark components are listed as follows:

- Spark SQL [16]: One important feature of Spark SQL is that it unifies the two abstractions: relational tables and RDD. So programmers can easily mix SQL commands to query external data sets with complex analytics. Concretely, users can run queries over both imported data from external sources (like Parquet files Hive Tables) and data stored in existing RDDs. In addition, Spark SQL allows writing RDDs out to Hive tables or Parquet files. It facilitates fast parallel processing of data queries over large distributed data sets for this purpose. It uses query languages called "HiveQL". For a fast application development, Spark develop the Catalyst framework. This one enables users via Spark SQL to rapidly add new optimizations.
- Spark streaming [17]: Spark Streaming is another component that provides automatic parallelization, as well as scalable and fault-tolerant streaming processing. It enables users to stream tasks by writing batches like processes in Java and Scala [18]. It is possible to integrate batch jobs and interactive queries. Spark streaming execute series of short batch jobs in-memory data stored on in RDDs is called as Streaming computation.
- MLlib [19]: Machine Learning Library (MLlib) is a Spark implementation of common Machine Learning functionalities related to binary classification, clustering, regression and collaborative filtering etc. to name a few.
- GraphX [20]: GraphX constitutes a library for manipulating graphs and executing the graph-parallel calculation. Similar to Spark SQL and Spark Streaming, GraphX enhance the significance of Spark RDD API. Hence clients can make a coordinated

Hadoop MapReduce: Data Sharing on Disk

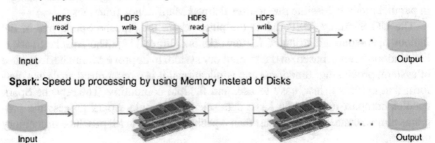

Spark: Speed up processing by using Memory instead of Disks

Fig. 2 Disk-based MapReduce versus Spark in-memory data processing

diagram with subjective properties connected to every vertex and edge. GraphX offers different operators that support graphs manipulation, for example, subgraph and Map Vertices.

The concept on which the Spark streaming operates is called as micro-batches. The Spark streaming accepts input dataset and divides that data into micro-batches [21], then the Spark engine processes those micro-batches to produce the final stream of results in sets/batches. Figure 2 shows the data processing details carried out by MapReduce through disk and the Spark processes the data in-memory, therefore the Spark is faster than MapReduce.

To Summarize

The Apache Spark is the most suitable platform for dynamic data/stream-data handling, and for real-time data analytics. In comparison with Hadoop, a Resilient Distributed Datasets (RDD) [15] is created and a Directed Acyclic Graph (DAG) [15] is prepared, as the related memory handling structures are maintained for Spark. It empowers the customers to store the data which is saved in the memory and to perform computation related to graph-cycle for comparable data, particularly from in-memory. The Spark framework saves enormous time of disk I/O operation. Hence Apache Spark is 100 times faster than Hadoop. The Spark can be set up to handle incoming real-time data and process it in micro-batches, and then save the result to resilient storage, such as HDFS, Cassandra, etc.

The Cassandra database is Distributed Database Management System, developed entirely in JAVA. The Cassandra has its own native query language called Cassandra Query Language, but it is a small subset of full SQL and does not support sufficient functionalities to handle aggregation and ad hoc queries.

3 Smart Cassandra Spark Integration (SCSI)

According to Mike Olson, Cloudera Chief Strategy Officer Sums up the breathtaking growth of the Spark and a paradigm shift in customer preferences witnessed by his Hadoop distributor company [22]—

> (Belissent, J.et al. 2010): *"Before very long, we expect that Spark will be the dominant general–purpose processing framework for Hadoop. If you want a good, general-purpose engine these days, you're choosing Apache Spark, not Apache MapReduce".*
>
> —Mike Olson

Not since peanut butter and jelly is an epic combo, the Spark is the world's foremost distributed analytics platform, delivering in-memory analytics with a speed and ease of use *unheard-of* in Hadoop. The Cassandra is the lighting fast distributed database powering IT giants such as Outbrain and Netflix.

Problem Statement

The Apache Spark is in-memory computing framework because it is preferably uses memory to cache partitions and having its own query language called as Spark SQL. The Apache Cassandra is Distributed NoSQL database. The Apache Cassandra is incompetent to handle aggregation and ad hoc queries. So when the Spark is paired with Cassandra, it offers a more *feature-rich* query language and allows to do data analytics that does not provided by Cassandra alone as well as the Spark alone.

The proposed framework *Smart Cassandra Spark Integration (SCSI)* is the collaboration of the Apache Cassandra and the Apache Spark. The proposed SCSI system architecture is as shown in Fig. 3 consist of a five key entities such as driver program, a cluster manager, workers, executors, and tasks. The master-slave model of distributed computing uses the SCSI. The node contains the cluster manager (or master process) is called master node of the cluster, while slave node contains the worker process. The functionalities or roles & responsibilities of a master node includes following:

- Resolve job details,
- Decide requirement of number of mappers and reducers for execution of jobs,
- Monitor the progress of every job,
- Keep log of every worker, etc.

The roles assigned to worker node or the worker nodes are conceptualized to gain following tasks:

- Receive, accept, and work on the commands from master node,
- Inform the status of every assigned work to the master,
- Execution of tasks and communication with local Cassandra database is the done by the executor,
- Entire application execution is done by the driver.

The proposed system SCSI uses the RDD data structures, which is called Resilient Distributed Datasets. RDD is dispersed datasets partitioned over processing nodes.

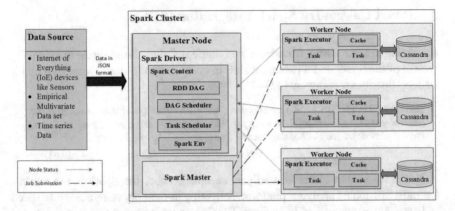

Fig. 3 High-level proposed SCSI system architecture

RDD records not only the number of partitions but also metadata of partitions. There are two ways to create the new RDD. The first, new RDD is generated whenever a transformation applies to an old RDD [23]. The recently created RDD object records, the dependency and transformation as a function object. The second new RDD is form by joining the various RDD together, the dependencies set of the new RDD is inherited from the parent RDD. The RDD transformation is same as declare a pointer to the particular object.

During the execution of a client application in SCSI system, the Driver and Worker plays an important role. The starting point for the execution of the application is the driver program, which performs the tasks scheduling, the tasks distribution, etc., over the multiple workers. Then workers manage parallel-processing tasks by using computing nodes and creates executor. The Executor communicates with the local data storage that means the Cassandra for task executions. The dataset for the system can be in the form of text, CSV, JSON, etc. For the proposed system JSON input file format is used. The sensors node generates the massive amount data in different formats, converts them into JSON formatto submit to the SCSI cluster. SCSI executes the task with the help of driver and worker, the details of execution are discussed in Sect. 3.2.

3.1 SCSI Node Deployment

This section of the chapter explains the role of SCSI nodes deployments on a standalone cluster. The new SCSI cluster is formed by an integration of the Apache Spark and the Apache Cassandra nodes in a big data cluster. Once the SCSI cluster is deployed, master nodes starts the master process and the slave node starts the worker process for controlling the entire cluster. Figure 4, demonstrates the integrated role of the Apache Spark and the Apache Cassandra nodes deployments on a standalone

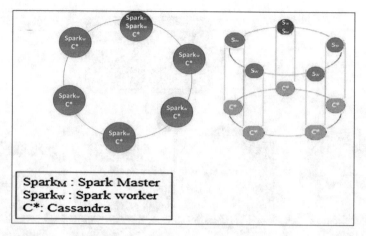

Fig. 4 The integrated role of the Apache Spark and Apache Cassandra nodes deployments on a standalone cluster

cluster. The standalone mode of cluster uses a basic cluster manager that is supplied with the Apache Spark. The spark master URL will be as follows:

SCSI://<hostname>:7077

Where, *<hostname>* is the name of the host to execute the master node details and *7077* is the port number, which is the default value, but it is configurable. This SCSI cluster manager supports First- In-First-Out (FIFO) job scheduling algorithm. It also allows concurrent application scheduling by setting the resource configuration options for each application. Each SCSI node in the cluster runs the Spark and the Cassandra combination. The logical flow of the execution of the application in the SCSI cluster is discussed in Sect. 3.2.

3.2 The Logical Flow of SCSI Cluster for Job Execution

Figure 5 describes the SCSI processes for execution of applications in parallel across the cluster nodes.

The Logical Flow (Algorithm) of Job Execution is as follows:

Step 1: Whenever sensor data is collected or client submits the application code, the Driver Program instantiates the Spark Context. The Spark Context converts the transformations and actions into logical DAG for execution. This logical DAG is then converted into a physical execution plan, which is then broken down into smaller physical execution units.

Step 2: The driver then interacts with the cluster manager or the Spark Master to negotiate the resources required to perform the tasks of the application code.

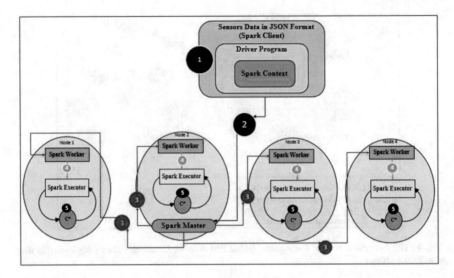

Fig. 5 The proposed sequential execution of client application

Step 3: The Spark Master then interacts with each of the worker nodes to understand the number of executors running in each of them to perform the tasks. The Spark worker consists of processes that can run in parallel to perform the tasks scheduled by the driver program, these processes are called the executors.

Step 4: The executors performs the data processing for the application code. In static allocation, the number of executors are fixed and run throughout the lifetime of the Spark application. In dynamic allocation, the user can also decide how many numbers of the executors are required to run the tasks, depending on the workload. Before the execution of tasks, the executors are registered with the driver program through the Master, so that the driver knows how many numbers of executors are running to perform the scheduled tasks. The executors then start executing the tasks scheduled by the worker nodes through the Spark Master

Step 5: Also, executor performs the following things:

i.   Read from and write the data to the Cassandra database as describe in Sect. 3.3.
ii. The prediction of sensor data stored in Cassandra (i.e., memory, or disk).

3.3   The Internals Details of SCSI Job Execution

Instead of using any cluster manager [4], SCSI comes with the facility of a single script that can be used to submit a program, called as *SCSI spark-submit*. It launches the application on the cluster. There are various options through which the *SCSI spark-submit* can connect to different cluster managers and controls many resources

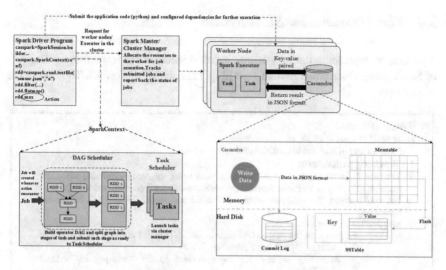

Fig. 6 The Internal details of the SCSI system during job execution

that application gets. For some cluster managers, *SCSI spark-submit* runs the driver within the cluster while for others, it runs only on the local machine. With this in mind, when *SCSI client submits* an application to the cluster with *SCSI spark-submit* this is what happens internally as shown in Fig. 6.

A standalone application starts and instantiates a SparkContext instance (and it is only then when the Spark client can call the application a driver). The driver program request to cluster manager or the Spark master for worker node/executor in the cluster, also asks for resources to launch executors. On behalf of the driver program, the cluster manager launches the executors. The driver process runs through the user application depending on the actions and transformations over RDDs task are sent to executors. The executors run the tasks and saves the result into the Cassandra. Whenever data is written into the Cassandra, it is first written in the Memtable (an actual memory) and at the same time, log files are committed to disk which ensures full data durability and safety. These log files are a kind of backup for each write to the Cassandra, which helps ensure data consistency even during a power failure because upon reboot, the data will be recovered in-memory from these log files. Adding more and more information to the Cassandra will finally result in reaching the memory limit. Then, the data stored by primary key is flushed into actual files on a disk called SSTables (sorted-strings table). Whenever any of the worker nodes does not respond to any query means that particular worker is not active, the tasks that are required to be performed will be automatically allocated to other worker nodes. SCSI automatically deals with this type of situation. For example, if Map () operation is executing on node1, and after some time node1, gets fail the SCSI rerun it on another node. Then execute the SparkContext. Stop () from the driver program or from the main method, all the executors terminated and the cluster resources released by the cluster manager.

3.4   The Algorithm for SCSI System

This section represents the lazy learning based solution for in-memory primitives of Spark using Distributed Matric Tree [24]. This algorithm forms a TopTree in the master node.

SCSI Algorithm steps:
1. **Input:** Dataset in JSON, nl
   //nl: Number of leaf tree to be distributed across the node
2. **Working procedure:**
   ➕ TopTree: build the top M-Tree in the master node using standard partitioning procedure and data sampling and same replicated with replication factor 1 on every worker node.
   ➕ For leaf node in the TopTree, subtree is created and stored as an RDD for further processing
   ➕ MapReduce element ε data
      ▪ Searches the nearest leaf node in the M-Tree and output a tuple with key (subtree ID) and value (element) [MAP]
      ▪ According to key, the tuple is sent to the corresponding subtree [SHUFFLE]
      ▪ The element with same key (subtree ID) are combine by inserting them into the local subtree [REDUCE]
   ➕ End MapReduce
3. **Output:** A tuple with element (key) and a list of neighbor for storage (value)

3.5   The List of Benefits of SCSI System

The SCSI system which is basically an innovative integration of the Apache Spark and the Cassandra proved various real-time system characteristics and they are as follows:

1. **Speedup**: It is observed that the speed of code execution by SCSI is hundred time faster than Hadoop MapReduce. This became feasible due to SCSI's supports to acyclic data flows and in-memory computations by using advanced Directed Acyclic Graph execution engine.
2. **Usability**: User is given freedom to code in varied high-level programming languages using Java, Scala, Python, and R, etc. SCSI has built user-friendly and compatible parallel apps from high-level operators to use it interactively with the Scala, Python, and R shells.

3. **Universal, broad, and comprehensive**: SCSI supports broad combination of SQL, real-time-dynamic data handling, data frames, MLlib, GraphX and the Spark functionalities together in single platform.

4. **Availability**: The varied range of platforms supported and integrated by SCSI includes standalone-SCSI with cluster-node in EC2 of Cloud Computing platform services, Hadoop, Mesos, etc. SCSI has the ability to passionately access

Table 1 Comparison of Cassandra queries with SCSI queries

Basis	Cassandra queries	SCSI queries
Join and union	No	Yes
Transformation	Limited	Yes
Outside data integration	No	Yes
Aggregations	Limited	Yes

data sources from HDFS, Cassandra, HBase, and S3. The SCSI can be executed using its standalone cluster mode, on EC2, on Hadoop YARN, and on Apache Mesos.

5. **Data Locality**: SCSI tasks are executed on the node that stores the data also. Hence its provide high data locality

6. **Execute the Aggregation Query**: SCSI system execute the SUM [25], MIN, MAX, AVG and other aggregations query. Also, execute the ad hoc queries. The Table 1 shows the comparison of Cassandra queries with SCSI queries.

3.6 Ancillary Learning About Integrated File Systems and Analytical Engine

i. **MapReduce and HDFS**

Few frameworks are very efficient for high-speed computation system being computation-intensive where some frameworks provide scalable data-intensive as well as computation-intensive architecture. The MapReduce has its traditional advantages with HDFS but with the growing necessity of real-time big data analytics, it needs to improve cluster resource management.

ii. **Spark and HDFS**

The Spark on HDFS is comparatively more suitable for big data analytics applications. The Spark supports in-memory computation features and the HDFS can deal with a huge volume of data. Together they provide high-speed processing with fault tolerance and data replication. The Spark's process the data by keeping intermediate results in main-memory (cache), because of this Spark is more suitable for those systems where iterative processing is required. Though the Spark on HDFS performs well for all analytical problems. They identify a correlation between different environmental indicators of the sensor datasets by using Hadoop and Spark.

iii. **MapReduce and Cassandra**

A new approach can be considered where Apache MapReduce will work on Cassandra data store, which reduces the read/write overhead. As the Cassandra is compatible with both the Apache MapReduce and the Apache Spark, its integration with the Hadoop MapReduce results in high fault tolerance.

Table 2 The existing technologies and their integration to support big data analytics

Storage	Data processing	Advantages	Disadvantages
HDFS	MapReduce/Yarn	• Scale-out architecture • Fault tolerance • Optimized scheduling • High availability	• Problem with resource utilization • Not suitable for real-time analytics
HDFS	Spark	• High computation speed • In-memory features • Data locality	• suitable for interactive processing
Cassandra	MapReduce/Yarn	• Scale-out architecture • Fault tolerance	• Not suitable for iterative processing
Cassandra	Spark	• Best suitable for iterative and interactive processing • High speed for parallel processing • Data locality • High scalability • Fault tolerance	• Complex structure

iv. **Spark and Cassandra**

For the real-time, online web and mobile applications dataset, the Apache Cassandra database is a perfect choice, whereas the Spark is the fastest processing of colder data in the data lakes, warehouses, etc. Its integration effectively supports the different analytic "tempos" needed to satisfy customer requirements and run the organization [26]. Their integration may result in high I/O throughput, the data availability, and the high-speed computation, the high-level data-intensive and the computation-intensive infrastructure as shown in Table 2.

4 The Mapreduce Streaming Above the Cassandra Dataset

4.1 The SCSI Streaming Pipeline over the Cassandra Datasets

Figure. 7 shows the details related to the SCSI proposal, introduced in this chapter. This section gives the details related to the SCSI pipelines for the Cassandra datasets. This SCSI streaming pipeline is designed using steps including Data preparation, transformation and processing. The details related to these three stages, performance

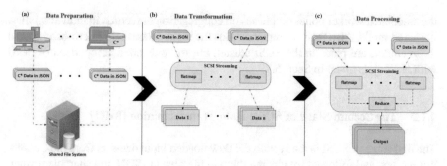

Fig. 7 The three stages of the SCSI streaming pipeline structure architecture

results, etc., are described in following Sect. 5. As shown in Fig. 7a shows, Data Preparation stage contains various worker nodes. These worker nodes are responsible for fetching the dataset from the local the Cassandra servers. Further, these worker nodes also carefully store the dataset details into the shared file system. Figure 7b shows, the Data Transformation stage which exports dataset in JSON format. The flat-map function is used to convert datasets into required specific formats. Figure 7c shows, the Data Processing stage details, which also uses the flat-map function. The task of this stage involves reducing the required procedures involved in reformatted data, produced in (b).

4.1.1 The First Stage of SCSI: Data Preparation

In the first stage of SCSI, for data preparation, the required input datasets are made available using the Cassandra servers. These datasets for ease of further processing are stored into the HDFS like distributed the file system having shared modes. As discussed by authors in [7], the Cassandra permits exporting data details from datasets fetched by servers into equivalent and required JSON formats. Each SCSI node is able to download data from linked-server powered by Cassandra to shared file formats, utilizing this built-in characteristic of the Cassandra. For every write request for the Cassandra server, data is first written in the Memtable (an actual memory) and at the same time, the log files are committed to the disk which ensures full data durability and safety. These log files are a kind of backup for each write to the Cassandra, which helps ensure the data consistency even during a power failure because upon reboot, the data will be recovered in-memory from these log files. Adding more and more information to the Cassandra will finally results in reaching the memory limit. Then, the data stored by primary key is flushed into actual files on a disk called Sorted-Strings Table, SSTables, (for details, refer the Sect. 3.3).

For the experimental setup of the SCSI framework, each worker is connected to its linked-server powered by the Cassandra and is capable to export actual Memory table called Memtable into the Stored String Table. Once the flushing of the data is completed, a worker starts the exporting operation. By using the "put" command, the associated worker nodes emphases on congregation of records in individual files, in shared mode. Also "put" command splits the input dataset into micro-batches and those chunks are place in the SCSI cluster. For more detailed comparison of SCSI with Hadoop is given in Sect. 5.1.

4.1.2 The Second Stage of SCSI: Data Transformation [MR1]

The first stage of SCSI is ready with the downloaded input datasets from the Cassandra servers, and followed by placing them to the files in JSON and sharable format. The SCSI architecture is proposed to handle issues related legacy application executables, which are difficult to rewrite or modify using Java for target results. The second stage of the SCSI architecture is Data Transformation (MR1) as shown in Fig. 7b.

The transformation phase involves in processing of each input records, and converts them to the required formats. The intermediate output files accommodate the results, as a part of transformation. This stage is used to start the flat-map operation of SCSI, however, no reduce operation is implemented yet. The responsibility of reducer function is to convert JSON files to appropriate format. The dependencies between nodes, data and or processing dependencies is not handled by this stage and hence best fitted for the SCSI streaming framework.

The Data Transformation stage of SCSI streaming is possible to implement in any programming language. For this research work the Python Scripts are used. The SCSI operations are based on the iterative data series, whose output becomes the input to the remaining stages of the SCSI streaming operations. This system allows users to specify the impactful attributes from given datasets and also convert the dataset in recommended file formats. This stage usually reduces data size, which ultimately improves the performance of the next stage. The SCSI and Hadoop streaming comparative analysis is further discussed in Sect. 5.2.

4.1.3 The Third Stage of SCSI: Data Processing [MR2]

The data processing stage execute the Python scripting programs, which was the initial target applications of data transformation, over the sensors data. The data is now available in a format that can be processed. In this stage of SCSI Streaming, flat-map and reduce operations is recommended and used, to run executables which is generated from the second stage of SCSI pipeline by using flat-map and reduce operation. The Hadoop and proposed SCSI streaming comparative analysis is further discussed in Sect. 5.3.

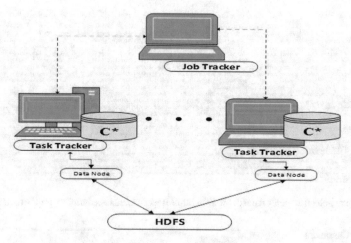

Fig. 8 The layout of Hadoop streaming over Cassandra dataset

4.2 Hadoop Streaming Pipeline Over Cassandra Datasets

The Hadoop streaming pipeline also works in three stages: data preparation, transformation, and processing. In the first stage, each data node of the Hadoop is connected to the linked-server powered by the Cassandra. By using the "put" command placed JSON formatted file into the Hadoop Distributed File System (HDFS) [11]. Also the "put" command splits the input dataset to the distribute data among the Data Nodes of the HDFS. These data-splits are input to the next stage (Data Transformation) of Hadoop Streaming pipeline. The HDFS requires the use of various APIs to interact with the files because it is a non-POSIX compliant file system [11]. The Hadoop streaming is implemented in non-Java programming language. The assumption is that the executables generated by the second stage of Hadoop streaming is not used by HDFS API due to this not having immediate access of the input splits. To address this issue, in the data processing, TaskTracker reads the input from the HDFS, processes them by data node and writes the results back to the HDFS. The layout of the Hadoop streaming over the Cassandra dataset is shown in Fig. 8.

5 Performance Results

This section explains the performance of the SCSI streaming system with the Hadoop streaming system with respect to computation speed. The experiments are carried out on a five nodes Spark cluster, one node is master, and the remaining four are workers. The Spark version 2.2.0 installed on each node, also installed the Apache Cassandra version 3.11.1 on each Spark slave nodes. The master node has Intel Core i5 6600 K 3.50 GHz Quad Core processor, 6 cores, a 64-bit version of Linux 2.6.15

	Date & Time	use [kW]	gen [kW]	House overall [kW]	Dishwash er [kW]	Furnace 1 [kW]	Furnace 2 [kW]	Home office [kW]	Fridge [kW]	Wine cellar [kW]	Garage door [kW]	Kitchen 12 [kW]	Kitchen 14 [kW]	Kitchen 38 [kW]	Barn [kW]	Well [kW]	Microwa ve [kW]	Living room [kW]	Solar [kW]
2	1/1/2016 0:00	0.932833	0.003483	0.932833	3.33E-05	0.0207	0.061917	0.442633	0.12415	0.006983	0.013083	0.000417	0.00015	0	0.03135	0.001017	0.004067	0.001517	0.003483
3	1/1/2016 0:01	0.934333	0.003467	0.934333	0	0.020717	0.063817	0.444067	0.124	0.006983	0.013117	0.000417	0.00015	0	0.0315	0.001017	0.004067	0.00165	0.003467
4	1/1/2016 0:02	0.931817	0.003467	0.931817	1.67E-05	0.0207	0.062317	0.446067	0.123533	0.006983	0.013083	0.000433	0.000167	1.67E-05	0.031517	0.001	0.004067	0.00165	0.003467
5	1/1/2016 0:03	1.02205	0.003483	1.02205	1.67E-05	0.1069	0.068517	0.446583	0.123133	0.006983	0.013	0.000433	0.000217	0	0.0315	0.001017	0.004067	0.001617	0.003483
6	1/1/2016 0:04	1.1394	0.003467	1.1394	0.000133	0.236933	0.063983	0.446533	0.12285	0.00685	0.012783	0.00045	0.000333	0	0.0315	0.001017	0.004067	0.001583	0.003467
7	1/1/2016 0:05	1.391867	0.003433	1.391867	0.000283	0.50325	0.063667	0.447033	0.1223	0.006717	0.012433	0.000483	0.000567	0	0.03145	0.001017	0.004067	0.001583	0.003433
8	1/1/2016 0:06	1.366217	0.00345	1.366217	0.000283	0.4994	0.063717	0.443267	0.12205	0.006733	0.012417	0.000517	0.00055	0	0.03155	0.001033	0.004117	0.001533	0.00345
9	1/1/2016 0:07	1.4319	0.003417	1.4319	0.00025	0.477867	0.178633	0.444283	0.1218	0.006783	0.01255	0.000483	0.00045	0	0.031733	0.001033	0.0042	0.00155	0.003417
10	1/1/2016 0:08	1.6273	0.003417	1.6273	0.000183	0.44765	0.3657	0.441467	0.121617	0.000695	0.012717	0.000467	0.0003	1.67E-05	0.031767	0.001017	0.0042	0.001567	0.003417
11	1/1/2016 0:09	1.735383	0.003417	1.735383	1.67E-05	0.17155	0.6825	0.438733	0.121633	0.007233	0.01335	0.000367	5.00E-05	0	0.031667	0.001017	0.0042	0.001617	0.003417
12	1/1/2016 0:10	1.585083	0.003417	1.585083	5.00E-05	0.0221	0.678733	0.4402	0.12145	0.007433	0.013583	0.00035	0.000117	3.33E-05	0.031667	0.001	0.0042	0.001567	0.003417
13	1/1/2016 0:11	1.510317	0.003433	1.510317	3.33E-05	0.021967	0.620667	0.43695	0.12125	0.007317	0.013533	0.000333	0.0001	0	0.03175	0.001	0.0042	0.001567	0.003433
14	1/1/2016 0:12	1.459867	0.00345	1.459867	5.00E-05	0.021883	0.577467	0.43995	0.121033	0.007233	0.013517	0.000367	8.33E-05	1.67E-05	0.031783	0.001	0.004217	0.001583	0.00345
15	1/1/2016 0:13	0.840583	0.003433	0.840583	0	0.02095	0.1448	0.444783	0.035017	0.007033	0.013183	0.000065	0.000183	0	0.031783	0.001017	0.004217	0.001617	0.003433
16	1/1/2016 0:14	0.7032	0.003433	0.7032	1.67E-05	0.020733	0.061967	0.443833	0.004783	0.006967	0.013117	0.000733	0.000233	0	0.03175	0.001017	0.004233	0.00155	0.003433
17	1/1/2016 0:15	0.571883	0.003433	0.571883	0	0.02065	0.06365	0.307783	0.004917	0.00705	0.0131	0.000783	0.00015	0	0.031733	0.001	0.004217	0.001583	0.00345
18	1/1/2016 0:16	0.485733	0.00345	0.485733	1.67E-05	0.020617	0.063433	0.22045	0.004983	0.00703	0.011117	0.00075	8.33E-05	0	0.031833	0.001	0.004233	0.001617	0.00345
19	1/1/2016 0:17	0.523167	0.003433	0.523167	0	0.020633	0.062117	0.26005	0.00495	0.007	0.013083	0.000733	0.0001	1.67E-05	0.03185	0.001017	0.00425	0.001717	0.003433
20	1/1/2016 0:18	0.5362	0.00345	0.5362	0	0.020683	0.062917	0.272067	0.00495	0.007033	0.01315	0.000733	0.000117	0	0.031867	0.001017	0.004233	0.00165	0.00345
21	1/1/2016 0:19	0.53415	0.00345	0.53415	1.67E-05	0.020667	0.06265	0.272067	0.00495	0.0071	0.01315	0.000733	0.0001	0	0.0319	0.001	0.004233	0.00175	0.00345

HomeC-meter1_2016

Fig. 9 Smart-electric meter dataset (*Source* http://traces.cs.umass.edu/index.php/Smart/Smart)

Table 3 Cassandra configuration

Parameter	Value
Cassandra heap size	2 GB
Partitioner	Random partitioner
Replication factor	1
Row caching	off

of 64-bit version and 32 GB of RAM. The slave node connected via 8 GB Gigabit Ethernet, which has Intel Xeon CPUs, 6 cores, 64 GB of RAM.

Dataset: The dataset to be used for research purpose is Electricity Smart Meter data connected at individual homes as shown in Fig. 9, total 50 K records in the dataset file. The case study also executed on different sensor data that means temperature, light and humidity datasets obtained from the UCI and other repositories [27].

For this research the Electricity Smart Meter dataset is chosen because next generation Smart-Energy-Meters is the necessity of this world today and it is the future. To effectually implement Smart-Energy-Meter systems on large scale, a true distributed system should be developed and maintained. Such Smart-Energy-Meters are the devices which are primarily based on IoT, GPS and Cloud computing concepts. The devices which are IoT enabled, generates massive amount of data, especially Smart-Energy-Meters will, for sure. To handle such multivariate massive data, SCSI is the best suitable model.

The Apache Cassandra and Apache spark configuration are specified in Tables 3 and 4 respectively. In addition to these, specified distributions of the default parameters shipped used by both Cassandra and spark.

Table 4 Spark configuration

Parameter	Value
dfs.block.size	128 MB
dfs.replication	1
Mapred.tasktracker.flapmap.tasks.maximum	6
Io.sort.mb	100 MB

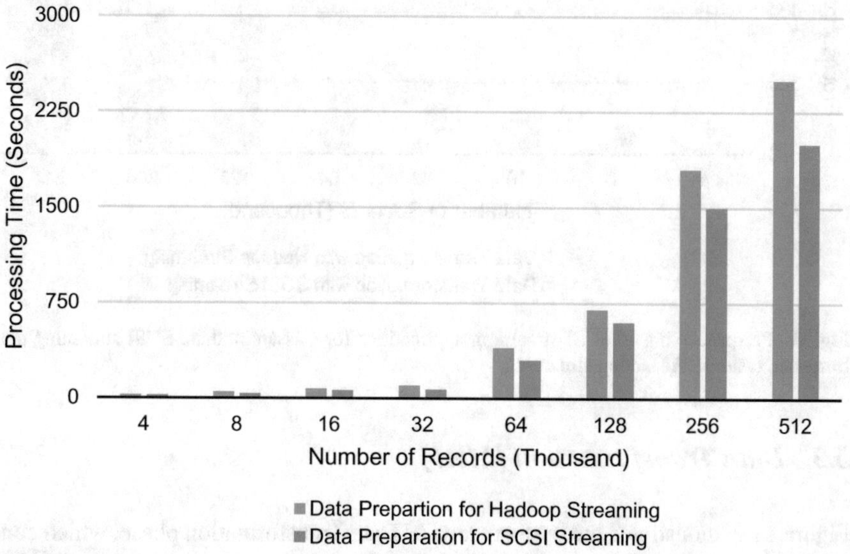

Fig. 10 The shifting of data from Cassandra server into the file system for SCSI and Hadoop Streaming. As the data sizes increases, cost is also increases

5.1 Data Preparation

For experimental setup, a worker node of the SCSI framework runs a Cassandra server. The aim is to store attributes like temperature, humidity, and light sensors data in the Cassandra. For this setup, replication factor is set to one for the Cassandra server's data distribution.

Figure 10 demonstrates the execution of SCSI streaming and Hadoop streaming by taking an image of the input dataset from the Cassandra into the shared file system for processing. As data size increases, the cost of shifting data from the Cassandra servers expectedly increases. There exists a linear relation between the shifting cost and the data size. The expense of shifting 4000 records takes nearly 60 times less than shifting 512 thousand input records. Additionally Fig. 10 demonstrates the difference between data preparation for SCSI Streaming and Hadoop Streaming. It is observed that at 4000 records, the speed of Data Preparation for SCSI Streaming is same as Hadoop streaming and it is more than 1.3 times faster at 64 and 256 thousand records.

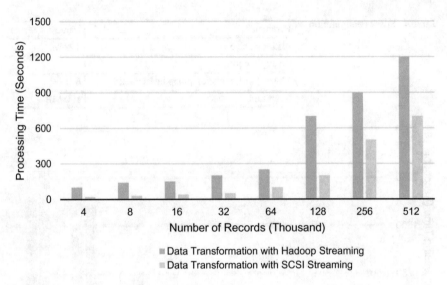

Fig. 11 Pre-processing of SCSI streaming applications for Cassandra data. SCSI streaming performance faster than Hadoop streaming

5.2 Data Transformation [MR1]

Figure 11 demonstrates the performance of Data Transformation phase, which converts the snapshot of the target dataset into the required format using both SCSI Streaming and Hadoop Streaming. At the 4000 input records, the speed of data transformation for SCSI streaming is ten times faster than Hadoop Streaming. At 64 and 256 thousand records SCSI streaming fifty percent faster than Hadoop streaming.

5.3 Data Processing [MR2]

This section explains the performance of executing the various application in Data Processing phase with the Hadoop Streaming and the SCSI Streaming. Also, it shows the overall cost for combining the data transformation and the data processing stages. As explained in Sect. 5.2, the Data Transformation stage not only converts the image of the input dataset into the desired format but also the size of the dataset is reduced. Due to the data size reduction, the data processing input to be much less as compared to the data transformation which ultimately improves the performance.

Figure 12 demonstrates the performance of the Data Processing, excluded in the Data Preparation and Data Transformation phases for the execution of the submitted application by user. At the stage of 4000 input records, the SCSI Streaming is 30% faster than the Spark Streaming and 80% speeder than the Hadoop Streaming. For the increased input data size, the Spark streaming is faster than the Hadoop Streaming

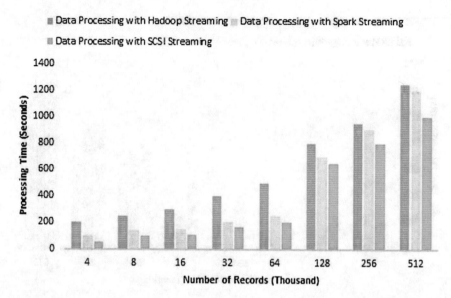

Fig. 12 Excluded the data preparation and data transformation phase during the data processing execution

and slower than the SCSI Streaming. At 64 and 256 thousand records, the speed of Data Processing for SCSI Streaming is fifty percent faster than Hadoop streaming.

Similarly, Fig. 13 demonstrates the performance of the execution of the same application, which includes Preparation and Transformation stages. For shifting the input dataset out of the database not only includes the cost of data movement but also needs split the exported data explicitly for each worker node in the cluster. At initial input data set, the SCSI Streaming is 1.3 times quick than the Hadoop streaming and at 512 thousand input records it is 3.6 times fast.

To Summarize

These two diagrams (Figs. 12 and 13) demonstrates that shifting information from the database accompanies a more cost which can easily overshadows the cost of accessing records straightforward from the database rather than the file system. As the research point of view, if the target application is advisable to keep running in data processing phase, it is best to store the input data in the cluster and use the SCSI system. That means reading the input by the Spark Mappers directly from the Cassandra servers and executes the given applications.

6 Related Work

As per Cluster of European research project on the Internet of Things [28], anything which having processing power is called as "Things", which are dynamic members

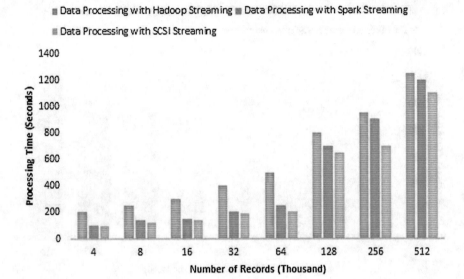

Fig. 13 Included the data preparation and data transformation phase during the data processing execution

in information and social processes, business and industries where they are enabled to interact and communicates with each other and with the environment by trading sensed data and information about the nature. IoT-based devices are capable of responding independently to the physical world's events and influencing it by running processes that trigger activities and make services with or without coordinate human intervention. As indicated by Forrester [22], a smart environment uses Information and Communications Technologies (ICT) to make the basic foundations, parts and administrations of a city organizations, medicinal services, public safety, real estate, education, transportation and utilities more aware, intuitive and effective. In [5, 29] explained the dataset generation technique by using the Horizontal Aggregation Method and by using the WEKA tool analyze this dataset. In [25] Structured Queries like Select, Project, Join, Aggregates, CASE, and PIVOT are explained.

There are plenty of applications which uses the NoSQL technology with MapReduce. In DataStax Enterprise [26] invented a system, in which Big Data [2] framework built on top of Cassandra which supports Hive, Pig, Apache Spark, and the Apache Hadoop. According to Thusoo et al. [29] Hive built on top of the Hadoop to support querying over the dataset which stored in a distributed file system like HDFS. According to the Huang et al. [23], Apache Spark is in-memory processing framework, also introduced the Remote Sensing (RS) algorithm and the remote sensing data incorporating with Resilient Distributed Datasets (RDDs) of Spark. According to Yang [30], Osprey is a middleware to provide MapReduce like adaptation to internal failure support to SQL databases. Osprey splits the structured queries into the number of subqueries. Also distributes the data over the cluster with replication factor three in classic MapReduce style. It provides the fault-tolerance and load balancing support to the Structured Query Language database, without concentrating on processing of data on such frameworks which utilizes the MapReduce style. According to the Kaldewey et al. [31], presents Clydesdale for handling organized information with Hadoop. They give a correlation about Hive, demonstrating the performance advantages of MapReduce model, more specifically Hadoop, but Clydesdale does not use the NoSQL database with non-Java applications.

7 Conclusion

To effectively combine the distributed data stores, such as Apache Cassandra with scalable distributed programming models, such as Apache Spark, it needs the software pipeline. This software pipeline allows users to write a Spark program in any language (Java, Scala, Python, and R) to make use of the NoSQL data storage framework concept. This research presented a novel scalable approache called *Smart Cassandra Spark Integration (SCSI)* for solving the challenges of integration of NoSQL data stores like Apache Cassandra with Apache Spark to manage distributed IoT data. This chapter depicts two diverse methodologies, one fast processing Spark working with the distributed Cassandra cluster directly to perform operations and the other exporting the dataset from the Cassandra database servers to the file system for further processing. Experimental results demonstrated the predominance and eminent qualities of SCSI streaming pipeline over the Hadoop streaming and also, exhibited the relevance of proposed SCSI Streaming under different platforms. The proposed framework is scalable, efficient, and accurate over a big stream of IoT data.

The Direction for future work is to integrate MPI/OpenMP with Cassandra. To improve speed performance of a Data Analytics in the Cloud, this idea is an interesting subject of the research.

Annexure

How to Install Spark with Cassandra
The following steps describe how to set up a server with both a Spark node and a Cassandra node (Spark and Cassandra will both be running on localhost). There are two ways for setting up a Spark and Cassandra server: if you have DataStax Enterprise [3] then you can simply install an Analytics Node and check off the box for Spark or, if you are using the open source version, then you will need to follow these steps.

This assumes you already have Cassandra setup.

1. Download and setup Spark

    i. Go to http://spark.apache.org/downloads.html.
   ii. Choose Spark version 2.2.0 and "Pre-built for Hadoop 2.4" then Direct Download. This will download an archive with the built binaries for Spark.
  iii. Extract this to a directory of your choosing: Ex. ~/apps/spark-1.2
  iv. Test Spark is working by opening the Shell

2. Test that Spark Works

    i. cd into the Spark directory
   ii. Run "./bin/spark-shell". This will open up the Spark interactive shell program
  iii. If everything worked it should display this prompt: "Scala>"
  iv. Run a simple calculation: Ex. sc.parallelize(1 to 100).sum(_+_)
   v. Exit the Spark shell with the command "exit"

The Spark Cassandra Connector

To connect Spark to a Cassandra cluster, the Cassandra Connector will need to be added to the Spark project. DataStax provides their own Cassandra Connector on GitHub and we will use that.

1. Clone the Spark Cassandra Connector repository: https://github.com/datastax/spark-cassandra-connector
2. cd into "spark-Cassandra-connector"
3. Build the Spark Cassandra Connector

    i. Execute the command "./sbt/sbt assembly"
   ii. This should output compiled jar files to the directory named "target". There will be two jar files, one for Scala and one for Java.
  iii. The jar we are interested in is "spark-cassandra-connector-assembly-1.1.1-SNAPSHOT.jar" the one for Scala.

4. Move the jar file into an easy-to-find directory: ~/apps/spark-1.2/jars

To Load the Connector into the Spark Shell

1. start the shell with this command:
../bin/spark-shell–jars~/apps/spark-1.2/jars/spark-cassandra-connector-assembly-1.1.1-SNAPSHOT.jar
2. Connect the Spark Context to the Cassandra cluster:

    i. Stop the default context: sc.stop
   ii. Import the necessary jar files:importcom.datastax.spark.connector._, import org.apache.spark.SparkContext, import org.apache.spark.SparkContext._, import org.apache.spark.SparkConf
  iii. Make a new SparkConf with the Cassandra connection details:Valcone=new SparkConf (true).set ("spark.cassandra.connection.host", "localhost")
  iv. Create a new Spark Context:valsc=new SparkContext(conf)

3. You now have a new SparkContext which is connected to your Cassandra cluster.
4. From the Spark Shell run the following commands:

    i. valtest_spark_rdd=sc.cassandraTable("test_spark", "test")
    ii. test_spark_rdd.first
    iii. The predicted output generated

References

1. Ray, P.: A survey of IoT cloud platforms. Future Comput. Inform. J. **1**(1–2), 35–46 (2016)
2. UMassTraceRepository. http://traces.cs.umass.edu/index.php/Smart/Smart
3. National energy research scientific computing center. http://www.nersc.gov
4. Apache Spark. http://spark.apache.org
5. Chaudhari, A.A., Khanuja, H.K.: Extended SQL aggregation for database. Int. J. Comput. Trends Technol. (IJCTT) **18**(6), 272–275 (2014)
6. Lakshman, A., Malik P.: Cassandra: structured storage system on a p2p network. In Proceeding of the 28th ACM Symposium Principles of Distributed Computing, New York, NY, USA, pp. 1–5 (2009)
7. Cassandra wiki, operations. http://wiki.apache.org/cassandra/Operations
8. Dede, E., Sendir, B., Kuzlu, P., Hartog, J., Govindaraju, M.: An evaluation of cassandra for Hadoop. In Proceedings of the IEEE 6th International Conference Cloud Computing, Washington, DC, USA, pp. 494–501 (2013)
9. Apache Hadoop. http://hadoop.apache.org
10. Premchaiswadi, W., Walisa, R., Sarayut, I., Nucharee, P.: Applying Hadoop's MapReduce framework on clustering the GPS signals through cloud computing. In: International Conference on High Performance Computing and Simulation (HPCS), pp. 644–649 (2013)
11. Dede, E., Sendir, B., Kuzlu, P., Weachock, J., Govindaraju, M., Ramakrishnan, L.: Processing Cassandra Datasets with Hadoop-Streaming Based Approaches. IEEE Trans. Server Comput. **9**(1), 46–58 (2016)
12. Acharjya, D., Ahmed, K.P.: A survey on big data analytics: challenges, open research issues and tools. Int. J. Adv. Comput. Sci. Appl. **7**, 511–518 (2016)
13. Karau, H.: Fast Data Processing with Spark. Packt Publishing Ltd. (2013)
14. Sakr, S.: Chapter 3: General-purpose big data processing systems. In: Big Data 2.0 Processing Systems. Springer, pp. 15–39 (2016)
15. Chen, J., Li, K., Tang, Z., Bilal, K.: A parallel random forest algorithm for big data in a Spark Cloud Computing environment. IEEE Trans. Parallel Distrib. Syst. **28**(4), 919–933 (2017)
16. Sakr, S.: Big data 2.0 processing systems: a survey. Springer Briefs in Computer Science (2016)
17. Azarmi, B.: Chapter 4: The big (data) problem. In: Scalable Big Data Architecture, Springer, pp. 1–16 (2016)
18. Scala programming language. http://www.scala-lang.org
19. Landset, S., Khoshgoftaar, T.M., Richter, A.N., Hasanin, T.: A survey of open source tools for machine learning with big data in the Hadoop ecosystem. J. Big Data 2.1 (2015)
20. Wadkar, S., Siddalingaiah, M.: Apache Ambari. In: Pro Apache Hadoop, pp. 399–401. Springer (2014)
21. Kalantari, A., Kamsin, A., Kamaruddin, H., Ebrahim, N., Ebrahimi, A., Shamshirband, S.: A bibliometric approach to tracking big data research trends. J. Big Data, 1–18 (2017)

Web References

22. Belissent, J.: Chapter 5: Getting clever about smart cities: new opportunities require new business models. Forrester Research (2010)
23. Huang, W., Meng, L., Zhang, D., Zhang, W.: In-memory parallel processing of massive remotely sensed data using an Apache Spark on Hadoop YARN model. IEEE J. Sel. Topics Appl. Earth Obs. Remote Sens. **10**(1), 3–19 (2017)
24. Soumaya, O., Mohamed, T., Soufiane, A., Abderrahmane, D., Mohamed, A.: Real-time data stream processing-challenges and perspectives. Int. J. Comput. Sci. Issues **14**(5), 6–12 (2017)
25. Chaudhari, A.A., Khanuja, H.K.: Database transformation to build data-set for data mining analysis—a review. In: 2015 International Conference on Computing Communication Control and Automation (IEEE Digital library), pp. 386–389 (2015)
26. DataStax Enterprise. http://www.datastax.com/what-we-offer/products-services/datastax-enterprise
27. Blake, C.L., Merz, C.J.: UCI repository of machine learning database. Department of Information and Computer Science, University of California, Irvine, CA (1998). http://www.ics.uci.edu/~mlearn/MLRepository.html
28. Sundmaeker, H., Guillemin, P., Friess, P., Woelfflé, S.: Vision and challenges for realizing the Internet of Things. In: CERP-IoT-Cluster of European Research Projects on the Internet of Things (2010)

Additional References

29. Thusoo, A., Sarma, J., Jain, N., Shao, Z., Chakka, P., Zhang, N., Antony, S., Liu, H., Murthy, R.: Hive-a petabyte scale data warehouse using Hadoop. In Proceedings of the IEEE 26th International Conference Data Engineering, pp. 996–1005 (2010)
30. Yang, C., Yen, C., Tan, C., Madden S.R.: Osprey: implementing MapReduce-style fault tolerance in a shared-nothing distributed database. In: Proceedings of the IEEE 26th International Conference on Data Engineering, pp. 657–668 (2010)
31. Kaldewey, T., Shekita, E.J., Tata, S.,: Clydesdale: structured data processing on MapReduce. In Proceedings of the 15th International Conference on Extending Database Technology, New York, NY, USA, pp. 15–25 (2012)

Erratum to: Machine Learning on Big Data: A Developmental Approach on Societal Applications

Le Hoang Son, Hrudaya Kumar Tripathy, Biswa Ranjan Acharya, Raghvendra Kumar and Jyotir Moy Chatterjee

Erratum to:
Chapter "Machine Learning on Big Data: A Developmental Approach on Societal Applications" in: M. Mittal et al. (eds.), *Big Data Processing Using Spark in Cloud*, **Studies in Big Data 43, https://doi.org/10.1007/978-981-13-0550-4_7**

In the original version of the book, the co-author name "Acharya Biswa Ranjan" in the chapter "Machine Learning on Big Data: A Developmental Approach on Societal Applications" has been changed as "Biswa Ranjan Acharya".

The updated online version of this chapter can be found at
https://doi.org/10.1007/978-981-13-0550-4_7

Printed in the United States
By Bookmasters